T-Labs Series in Telecommunication Services

It is the aim of the Springer Series in Telecommunication Services to foster an interdisciplinary exchange of knowledge addressing all topics which are essential for developing high-quality and highly usable telecommunication services. This includes basic concepts of underlying technologies, distribution networks, architectures and platforms for service design, deployment and adaptation, as well as the users' perception of telecommunication services. By taking a vertical perspective over all these steps, we aim to provide the scientific bases for the development and continuous evaluation of innovative services which provide a better value for their users. In fact, the human-centric design of high-quality telecommunication services – the so called "quality engineering" – forms an essential topic of this series, as it will ultimately lead to better user experience and acceptance. The series is directed towards both scientists and practitioners from all related disciplines and industries.

** Indexing: books in this series are indexing in Scopus **

Robert Philipp Spang

Individualized Quality of Experience Estimation in Audiovisual Communication

Springer

Robert Philipp Spang
Quality and Usability Lab
Technische Universität Berlin
Berlin, Berlin, Germany

ISSN 2192-2810 ISSN 2192-2829 (electronic)
T-Labs Series in Telecommunication Services
ISBN 978-3-031-67550-8 ISBN 978-3-031-67551-5 (eBook)
https://doi.org/10.1007/978-3-031-67551-5

This Springer imprint is published by the registered company Springer Nature Switzerland AG
The registered company address is: Gewerbestrasse 11, 6330 Cham, Switzerland

The purpose of life is to live it, to taste experience to the utmost, to reach out eagerly and without fear for newer and richer experience.
—Eleanor Roosevelt

Preface

This book explores the impact of individuality on Quality of Experience (QoE) ratings by introducing and evaluating the Individuality Assisted Estimation (IAE) approach. Individuality, in this context, refers to the unique combination of characteristics, traits, behaviors, and qualities that differentiate one person from another. It recognizes that individuals possess their own set of personal attributes, preferences, and experiences, which contribute to their perception and rating of multimedia quality.

Traditional methods for assessing audiovisual quality, while accurate, overlook the influence of individual differences, neglecting the potential explanatory value that knowledge about an individual brings to the estimation process. This motivated the development of the IAE model, which integrates psychological traits and states into the QoE estimation process, aiming to enhance the accuracy of QoE ratings by tailoring results to individuals rather than relying on a Mean Opinion Score (MOS).

A comprehensive investigation was conducted to identify the most significant factors contributing to QoE ratings. The evaluation of the proposed model involved a dual-method approach, utilizing a video-based telephony simulation in a controlled environment and real-life audiovisual conversational scenarios. Comparing these two testing methods provided an evaluation of the simulation's effectiveness in studying human reactions to quality degradation.

The primary objective of this work was to refine QoE assessment methods by considering individual differences. The success of the IAE model was measured by its ability to QoE rating estimations compared to non-individuality-aware procedures. The research presented evidence supporting the effectiveness of the IAE, although the improvement over signal-based estimations was marginal.

Specifically, the IAE approach demonstrated significant efficacy in estimating task load, outperforming the signal-based model. This finding suggests potential avenues for targeted application of the IAE. The research also identified areas for further investigation and refinement, such as exploring the shared variance between individuality data and signal-based estimators.

In conclusion, the findings shed light on the crucial role of individual differences in QoE assessment and offer a validated and practical tool for researchers and practitioners. This book serves as a stepping-stone toward a deeper understanding of QoE, establishing a strong foundation for future research on this multifaceted concept.

Berlin, Germany

Robert Philipp Spang

Acknowledgments

Over the past years, working at the Quality & Usability Lab, I've had the exceptional privilege of knowing there is no place I'd rather be. It's clear to me that this sentiment is far from ordinary, but rather, it is the product of an extraordinarily welcoming, supportive, and open team, with whom I found joy in collaboration on many occasions. My deepest gratitude goes to my cherished friends and colleagues who made this journey unforgettable: Sebastian Möller, Irene Hube-Achter, Yasmin Hillebrenner, Wafaa Wardah, Thilo Michael, Vera Schmidt, Jan-Niklas Voigt-Antons, Kerstin Pieper, Martin Josef Burghart, Leon Schreiber, Jenny Bungart, Tanja Kojic, Maurizio Vergari, Max Warsinke, Philine Görzig, Tim Polzehl, Salar Mohtaj, Charlott Jakob, Clemens Feldhaus, Stefan Hillmann, Neha Deshpande, Sai Sirisha Rallabandi, Michael Wagner, Tobias Jettkowski, Daniel Conde, Razieh Khamseh-Ashari, Philipp Harnisch, Navid Ashrafi, Adnan Ahmad, Purna Dutta, Carola Trahms, Gabriel Mittag, Steven Schmidt, Saman Zadtootaghaj, Babak Naderi, Britta Hesse, Patrick Ehrenbrink, Tilo Himmelsbach, Neslihan Iskender, Luis Meier, Maria Fleßner-Jung, Andres Pinilla Palacios, Falk Schiffner, and Vinicius Woloszyn.

My thanks extend to the experts and mentors who graciously accepted this considerable commitment: Prof. Dr-Ing. Sebastian Möller, Prof. Dr. Mylène C.Q. Farias, Prof. Dr. Peter Reichl, and Prof. Dr. Henning Sprekeler.

Additionally, I am thankful to all the B.Sc. and M.Sc. students I had the honor of supervising throughout their thesis-writing processes. Your unique experiences and perspectives have been instrumental in shaping me as an educator. While this is a mission that will never be over, I am profoundly grateful for every shared moment working with: Rudresh, Anke, Oli, Arik, George, Michi, Anna, Max, Nebras, Barthi, Ben, Luisa, Erik, Leonie, and Leon. My appreciation extends to each student involved in the courses I had the opportunity to support or lead.

I've been fortunate to collaborate with some truly incredible individuals who, among other invaluable lessons, fostered my enthusiasm and curiosity as a researcher: Stefan Brandenburg, Jens Berger, Julie O'Sullivan-Solte, Paul Gellert, Milos Cernak, Pablo Minar, and Karl El Hajal, Esther Kim, Elizabeth Rochon, Laura

Laird, Raimund Schatz, and Philip Mildner, Sandra Mümken, Christine Haeger, Sonia Lech, and Johanna Schuster.

I owe a special debt of gratitude to those who set an admirable example of effective supervision during my own B.Sc. and M.Sc. theses. You were the catalysts that kindled my passion for research and offered consistent encouragement throughout my journey: Gemma Learmonth, Franziska Horn, Stefan Brandenburg, Nils Backhaus, Anna Trukenbrod, and Manfred Thüring—for your exemplary supervisory skills and your ever-present readiness to engage in discussions on the fundamental practices of science.

Last but far from least, my deepest thanks go to my family, Klaus, Jutta, and above all, to Charlotte. Your unwavering support and love have been the bedrock upon which my academic journey has been built. Thank you. For everything.

Contents

Acronyms

ACR	Absolute Category Rating
AVE	Average Variance Extracted
BH	Bonferroni–Holm
BIC	Bayesian Information Criterion
CX	Customer Experience
DOVER	Disentangled Objective Video Quality Evaluator
DSMM	Differential Susceptibility of Media Effects Model
ECG	Electrocardiogram
EDA	Electrodermal activity
EEG	Electroencephalography
EMG	Electromyography
FDS	Frustration Discomfort Scale
GBR	Gradient Boosting Regressor
HRV	Heart Rate Variability
HR	Heart Rate
HTMT	Heterotrait–Monotrait
IAE	Individuality Assisted Estimation
ITU	International Telecommunication Union
MAE	Mean Average Error
ML	Machine Learning
MOS	Mean Opinion Score
NISQA	Non-Intrusive Speech Quality Assessment
OSF	Open Science Framework
OTT	Over-the-Top
PESQ	Perceptual Evaluation of Speech Quality
PLCC	Pearson's Linear Correlation Coefficient
PLS	Partial Least Squares
POLQA	Perceptual Objective Listening Quality Assessment
PPG	Photoplethysmographie
QoE	Quality of Experience
QoS	Quality of Service

RBF	Radial Basis Function
RMSE	Root Mean Square Error
RMSSD	Root Mean Square of Successive RR-Interval Differences
RRI	RR-Intervals
SCL	Skin Conductance Level
SCR	Skin Conductance Response
SCT	Short Conversation Test
SEM	Structural Equation Model
SNS	Sympathetic Nervous System
SROCC	Spearman's Rank Order Correlation Coefficient
SVR	Support Vector Regressor
TLX	NASA Task Load Index
UB	User Behavior
UGC	User Generated Content
UX	User Experience
VQA	Video Quality Assessment

Chapter 1
Introduction

Research is to see what everybody else has seen, and to think what nobody else has thought.
Albert Szent-Györgyi

Video telephony, the art of transmitting and receiving real-time visual and auditory signals for the purpose of communication, has long captivated the human imagination. A century ago, visionaries and futurists dreamt of a world where individuals could converse face to face without the constraints of physical distance, envisioning the profound impact such an innovation would have on social and professional interactions, compare Fig. 1.1. Today, video telephony is not only a reality but a crucial and fascinating technology that underpins myriad aspects of modern life, from casual conversations with loved ones to high-stakes business meetings. As we increasingly rely on this mode of communication, it becomes essential to research and understand the factors that shape our experiences and impressions, such as the quality of the connection, conversation, image, and speech signal. This book seeks to delve into the realm of Quality of Experience (QoE) and examine the influence of individual and personal factors on the efficacy and satisfaction derived from video telephony, a technology that continues to revolutionize human interaction.

This chapter serves as the initial introduction and sets the tone for the entire work. It highlights the significance of the chosen topic and why it is intriguing to explore. Additionally, the chapter outlines the fundamental approach and concludes by posing the research questions that will guide the investigation.

1.1 Motivation

Multimedia quality perception is commonly measured by means of a mean opinion score (MOS). This score forms an average rating of how a number of typical users would rate their quality perception under standardized, typically neutral conditions. It may be predictive of how other users of the target group would rate their

R. P. Spang, *Individualized Quality of Experience Estimation in Audiovisual Communication*, T-Labs Series in Telecommunication Services,
https://doi.org/10.1007/978-3-031-67551-5_1

Fig. 1.1 The "correspondence cinema," imagined by French artist Jean-Marc Cote in 1910. The picture shows a woman and a man corresponding via in an imagined video-telephony apparatus. It appeared in a series of predictions about the future and about how people more than hundred years ago imagined the technology of our time. (Image credit: Wikimedia Commons)

perception when being confronted with the same stimuli. The score may also be estimated (predicted) through parameters of its transmission, a signal analysis, or a combination of both. Such data, for example, bandwidth, packet loss, or, e.g. various resolution aspects (e.g., color, temporal, or spatial resolution), provide technical details about the content's quality. While this yields good results for an average user group, this approach does not consider the user as an individual, with specific needs, preferences, and previously made experiences. As a consequence, today's signal-based quality estimations only provide an estimation of an average perception but do not predict an individual's perception, as they lack information about that individual.

The relevance of considering individual user perceptions in multimedia quality assessment lies in the fact that users are not a homogeneous group, but rather a diverse collection of individuals with unique needs, preferences, and experiences. By relying solely on mean opinion scores (MOS) and signal-based quality estimations, the nuances and variations in individual quality perception are not fully captured. As a result, the development of multimedia services, applications, and products may not cater for the specific requirements of each user, leading to suboptimal user experiences and potential dissatisfaction.

By acknowledging and accounting for individual user perceptions, developers and service providers can create more personalized experiences, enhancing satisfaction, usability, and overall enjoyment. This approach is particularly relevant in today's world, where user-centric design and personalization have become key factors in product success and user retention. Furthermore, understanding individual quality perception can help identify areas for improvement, inform adaptive streaming algorithms, and assist in developing technologies that better serve the diverse needs of users. Ultimately, considering individual user perceptions in multimedia quality assessment is essential for creating more inclusive, accessible, and satisfying experiences in an increasingly interconnected and digital world.

I seek to contribute and discuss possible ways toward an individualization of QoE estimations, reflecting characteristics of the individual in addition to group-specific characteristics. For this goal, I review possible individual characteristics, which might play a role for quality perception and judgment, taking video telephony as an exemplary service. While I believe the individualization of experience estimations of other services, for example, in the gaming context or the passive listening or viewing context, will follow identical considerations and processes, I focus on the video-telephony setting in the context of this work. Video-telephony is a bidirectional, multimodal way of communication and therefore covers many aspects involved in multimedia consumption and interaction.

This work contains broad considerations of the most discussed dimensions of the literature and substantiates these theories with empirical evidence from multiple experiments and two experimental paradigms to investigate overarching relationships and their predictive values. This adds a new model of QoE-formation and evidence-based conclusions to the field of engineering, the academic discipline that typically engages with quality and usability engineering. These additional aspects are borrowed from psychology, cognitive neuroscience, and biophysiology. Therefore, this interdisciplinary work first outlines the current state of the art of previous research. Following, I formulate the research questions I seek to answer throughout this book. Hereafter, I outline the structure of the manuscript to guide you through the thought processes, experiments, and interpretations of the collected data.

1.2 Book Structure

After this Chap. 1 motivated the general research area and lays out the structure of the rest of this manuscript, Chap. 2 of this book details the related work that is needed for the newly developed model of QoE and its estimation approach. It introduces the main concepts of individuality, its dimensions, and relevant concepts. It gives an overview of the available standards and theories of QoE, psychological foundations, and psychophysiological concepts of individuality, traits, and states. It thereby leads to the missing space in the literature that this book fills. Hence, the *Fundamentals*, Chap. 2 conclude with the research questions of this work.

Chapter 3 then introduces the Individuality Assisted Estimation (IAE) model of experience, which contains the theory development for this work. The model builds relationships with concepts of the QoE-domain, cognitive psychology, and

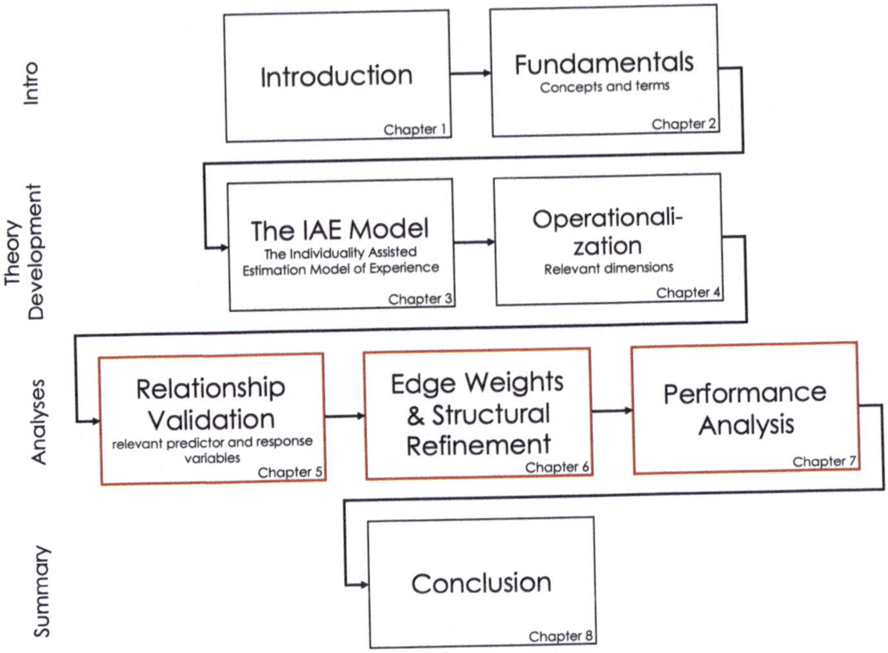

Fig. 1.2 Illustration outlining the organization of chapters within this book

psychophysiology, and explains the processes and connections that from a user's experience. It provides a rational what to include and measure to capture a person's individuality and how to integrate it for improved experience estimations. Chapter 4 continues in detail what the model suggested generally. Here, specific dimensions, measures, tools, and procedures are explained that are used in this work to quantify the previously introduced individuality elements.

Following, Chaps. 5, 6, and 7 build the core of this work through empirical analyses of the proposed model and its assumptions. To thoroughly test the IAE model, first, in Chap. 5, all proposed relationships are validated through two different experimental paradigms. Chapter 6 discusses a statistical model to evaluate the entire IAE structure all at once. This allows to quantify the relative importance of each of the model's components and to discuss what parts show the strongest links between a person's QoE ratings, the individuality of the person, and the stimulus at question. Lastly, Chap. 7 showcases an application of the model and enables more precise QoE estimations through the inclusion of individuality metrics. This triad of evaluation chapters dissects the model's assumptions, ranks the influence of its various elements, and showcases its predictive value.

Finally, Chap. 8 recapitulates the research questions, all main results, and discusses their implications. The chapter also provides an outlook on future work. Figure 1.2 visualizes this structure and how the chapters build four logical units.

Chapter 2
Fundamentals

If I have seen further it is by standing on the shoulders of Giants.
Sir Isaac Newton

This chapter lays the foundation for the entire thesis, providing essential explanations of key terms, concepts, and models. It also explores the existing literature on these elements and their effectiveness in estimating perceived quality of experience. In the latter part of the chapter, I spotlight the research questions, anchoring them to the broader scholarly discourse on QoE.

2.1 Quality, Experience, and Quality of Experience

Quality, an abstract construct of the mind, varies slightly in definition depending on the context in which it is used. At its core, the term represents a measure of excellence, or a state that is free from defects, deficiencies, and significant variations. According to an early ISO definition, quality is focused on customer satisfaction and is described as "the ability of a set of inherent characteristics of a product, system, or process to fulfill requirements of customers and other interested parties" (ISO Standard 9000, 2000).

This broad term can be applied to any product, service, or concept to determine its superiority or grade. Key elements that determine quality include functionality, durability, reliability, consistency, and overall user satisfaction. These facets of quality are frequently managed through quality assurance and quality control in various industries.

Experience in a user or customer context pertains to the overall interaction and engagement a person has with a product, service, or organization over time. This includes a wide spectrum of responses—emotional, physical, intellectual, and sensory.

R. P. Spang, *Individualized Quality of Experience Estimation in Audiovisual Communication*, T-Labs Series in Telecommunication Services,
https://doi.org/10.1007/978-3-031-67551-5_2

QoE interweaves the concepts of quality and experience, referring to a user's degree of delight or annoyance with an application or service. It emerges from the fulfillment of a user's expectations about the utility and/or enjoyment of a product or service, considering the user's personality and current state (Qualinet, 2013). Importantly, QoE is more than just a measure of a product or service's functionality or performance; it represents the entirety of the user's interaction and overall experience. As such, it is subjective, highly personal, and influenced by a range of factors including user expectations, context, and personal preferences.

Another key consideration in the realm of QoE is its distinction from Quality of Service (QoS). While QoS focuses on technical factors, QoE centers on the end-user experience, blending objective measures of user performance with quantitative subjective measures. Though QoS is crucial for the smooth functioning of a network service, it does not inherently guarantee a high QoE. A service may excel in terms of bandwidth, delay, and packet loss, but if it fails to meet user expectations or usability requirements, the overall QoE may still be poor (Brooks & Hestnes, 2010). Thus, it's critical to separately measure QoS and QoE.

Recent studies, such as Schmitt et al. (2017), emphasize the importance of considering individual user preferences and the interaction context when evaluating QoE in multiparty videoconferencing. Their empirical investigation using a Lego building task highlighted how variations in video quality, influenced by factors like encoding bitrate and packet loss, impact user interaction and perceived experience.

It is important to recognize that our experiences are internal processes that can only be accessed through introspection. These experiences are influenced by various factors, including our emotional states (Frijda, 1994; Forgas & Bower, 1987; Angrilli et al., 1997). This emphasizes the subjective nature of QoE, which aims to assess the overall acceptability of an application's quality from the perspective of end-users.

While the measurement of QoE typically involves gathering ratings from human participants, software systems are employed to estimate such ratings based on diverse inputs, as discussed in the subsequent sections. Examples of such estimation systems include Perceptual Evaluation of Speech Quality (PESQ) and the E-model (Pornpongtechavanich & Daengsi, 2019).

2.1.1 Conceptual Model of Quality of Experience

Numerous functional models of internal information processing in the context of quality assessment have been developed (Qualinet, 2013; Jekosch, 2005; Raake, 2007; Raake & Egger, 2014; Côté, 2011). Uhrig suggests to divide these into two categories: "psychological" and "psychophysiological". Psychological models incorporate solely psychological constructs, while psychophysiological models also include physiological constructs (Uhrig, 2021).

The internal quality formation conceptual model proposed by Raake and Egger (2014) presents a comprehensive integration of previous model versions. This model is designed to encapsulate the experience of an individual exposed to multimedia signals, who subsequently evaluates and describes their subjective quality experience. However, it's worth noting that this model is designed to describe passive reception of signals, thus omitting internal processing needed for human interactions. It also doesn't incorporate a system part that might provide details about system properties or parameters influencing the user's experience.

The model by Raake and Egger delineates internal quality formation into three phases: Perception, Evaluation, and Encoding. Each phase includes specific processes, representations, and their interrelations along with experiential correlates. The model also takes into account the user's emotional and motivational states, expectations based on prior knowledge, attitudes, preferences, and task-related information.

The Perception phase commences with signals from sensory modalities, primarily vision and audition, along with contextual information entering the sensory processing stage. This stage generates a multidimensional topographic representation of sensory features (Raake & Blauert, 2013). The Perception phase further involves steps of perceptual organization, leading to the formation of a perceptual event (Raake & Blauert, 2013; Bregman, 1994; Moore, 2012). The perceptual event is compared against perceptual references and expectations, enabling object recognition.

Quality awareness can be heightened due to a mismatch between the perceptual event and expectations, leading to exploratory motor responses. Alternatively, quality awareness can be influenced by conspicuous degradation in perceptual quality features or task-specific requests for quality ratings.

The Evaluation phase begins once quality awareness triggers cognitive quality evaluation. This phase starts with reflecting upon the perceived quality features compared to the expected or desired ones. A judgment process takes place, comparing the perceived and expected quality features, culminating in a quality event (Uhrig, 2021).

The Encoding phase typically occurs during quality tests. Here, the quality event is explicitly assigned to a certain category within a chosen sign system, like a numeric value or semantic label on a rating scale. This process results in a behaviorally manifested encoded quality event such as a quality rating.

During interactive situations involving multiple participants, additional perceptual and cognitive processes occur. An adapted version of the internal quality formation model for multiparty teleconferencing services has been proposed by Skowronek and Raake (2015). It introduces a new process, quality-attention focus, denoting the attention paid to a limited number of quality features. Factors related to selective attention are integral to participants' quality descriptions.

Even without the task-driven intention to evaluate and describe their experienced quality, human perceivers may focus on perceptual quality features due to salient changes. Such situations may capture attention and trigger orienting responses without necessarily involving higher-cognitive processing (Uhrig, 2021).

Recent attempts have been made to incorporate physiological phenomena into these models as indicators of internal processes and representations (Antons,

2015; Arndt, 2016). A detailed "psychophysiological" model of quality perception, founded on the model by Raake and Egger and further developments by Arndt and Antons, has been proposed by Uhrig (2021).

The models discussed thus far incorporate user states and various human influence factors; however, they do not specifically target the individuality of the user. Reichl et al. (2015) introduced a framework for comprehending QoE and User Behavior (UB) by explicitly modeling user states. This framework contemplates three distinct perspectives: the technical viewpoint, highlighting system factors impacting QoE; the eudaimonical viewpoint, concerning user well-being; and the behavioral viewpoint, of interest to service providers.

It pursues a cascading approach where system performance affects user state, subsequently influencing user behavior. The user state includes a range of factors affecting QoE and user behavior, such as perception processes, prior experiences, demographics, and usability.

Schmitt et al. (2017) addressed this gap by proposing models that incorporate user factors such as prior experience with technology and current affective state, significantly improving the accuracy of QoE predictions for multiparty videoconferencing scenarios.

The factors impacting QoE can be broadly classified into three primary categories: technical, user, and context. Technical influences encompass network and application-level parameters capable of affecting QoE, such as network delivery bandwidth, page load time, and video codec. User influences involve demographic, socioeconomic, and emotional characteristics, as well as the user's knowledge, preferences, and attitudes, all of which can directly alter user behavior and indirectly impact QoE. Contextual influences refer to the physical and social environments in which services and devices are used, including factors like location, social setting, and time pressure during an action. These factors can shape behavioral patterns and, consequently, influence QoE.

This framework by Reichl et al. (2015) could potentially pave the way for deriving a QoE model that considers user states, behavior, and technical aspects to formulate an individualized quality rating estimation, and the book lists several interesting factors to consider for such an approach. These ideas will be picked up in the following two chapters of this thesis.

2.1.2 Methods of Quality Assessment in the (Video-) Telecommunication Context

The evaluation of QoE can be complex, requiring a blend of both subjective and objective assessment methods. Here, we discuss various approaches and methods that have been proposed for assessing the QoE in this context.

Starting with a structured approach to quality assessment, Möller et al. (2009) introduced a comprehensive "Taxonomy of QoS and QoE Aspects for Multimodal

Human-Machine Interactions." This taxonomy is structured into three layers: QoS-influencing factors related to the user, system, and usage context; QoS interaction performance aspects that describe user-system behavior and performance; and QoE aspects tied to the internal quality perception and judgment processes of the user. The taxonomy serves as a practical guide to amass prior results on QoE assessment in video telephony, offering influencing factors for a user's quality rating, and thus supplying a set of attributes to consider when modeling user experience and quality rating formation. This taxonomy further differentiates between static user characteristics (like age, gender, native language) and dynamic ones (such as motivation or emotional state) (Wechsung et al., 2012).

Turning to quality assessment methods, the first to discuss is the traditional user perception approach. Historically, this method primarily relied on collecting subjective user data about media quality perception (Brooks & Hestnes, 2010). However, its primary limitation is the exclusive focus on the perceptual domain, overlooking user behaviors and interactions that occur post-perception.

An alternative method that broadens the focus is the User Experience (UX) approach. The UX approach scrutinizes objective user performance measures, including the correct recognition of visual information, the degree of problem-solving success, and conversational turn-taking dynamics. By aligning this approach with traditional user perception and QoE measures, a more comprehensive UX assessment can be achieved, encompassing both subjective and objective measures. This includes collecting quantitative data on subjective measures and employing statistical techniques to synthesize multiple variables into a holistic QoE measure (Brooks & Hestnes, 2010).

In their work, Farias et al. (2005) put forth a No-Reference approach for video quality assessment utilizing data hiding. This approach involves inconspicuously incorporating a signal into the source content, which is then analyzed post-transmission. By examining the retrieved signal, which may have undergone degradation, it's possible to estimate the quality of the video.

Further drilling down to specific techniques, rating paradigms such as the Absolute Category Rating (ACR), ACR with Hidden Reference (ACR-HR), Degradation Category Rating (DCR), and Pair Comparison (PC) method are used. These approaches rely on subjects' opinions, assigning mean opinion scores to evaluate the overall quality of a speech or video artifact.

Each of these methods has its strengths. For instance, ACR is efficient and productive, suitable for qualification tests. ACR-HR helps reduce biases inherent in the reference video and monitor. DCR is valuable for detecting transmission fidelity, color impairments, and skipping errors, while the PC method excels when test items are almost equivalent in quality.

The standard protocol for conducting an audiovisual conversation test comprises the procedures delineated in ITU-T Recommendation P.805 (2007) for phone-based dialogs and the methodologies outlined in ITU-T Recommendation P.920 (2000) for audiovisual communication. The latter resource offers precise guidelines for organizing tests, determining stimuli, managing participants, and accounting for environmental and equipment factors integral to audiovisual communication trials.

These experimental procedures outlined in ITU-T Recommendation P.920 (2000) are designed around task-oriented conversational situations for video-telephony applications. It also provides guidelines for devising new situations that ensure ecological validity based on the purpose of the assessment. An important aspect of the process is that the quality assessment happens post-conversation. Participants first rate the combined audiovisual quality, followed by individual ratings for video and then speech quality. This structure prevents the overall quality rating from being a simple average of the speech and video scores. Consequently, a single quality rating for speech, video, and audiovisual domains is obtained per experiment.

These tests are typically carried out in a controlled lab environment, with two participants positioned in separate, acoustically isolated rooms. They communicate over a real-time video-telephony system following predetermined conversational scenarios. Each interaction, typically lasting no more than 5 minutes, concludes with the participants assessing the quality of their connection using a five-point ACR scale, spanning from "Excellent" to "Bad" (Hammer, 2006).

2.1.2.1 Speech Quality, Video Quality, Audiovisual Quality, and Conversational Quality

Based on the aforementioned ISO definition of the quality term (ISO Standard 9000, 2000), Jekosch (2005) tailored the definition for speech quality assessment, stating that quality arises from "the judgment of the perceived composition of an entity with regard to its desired composition."

The "perceived composition" refers to the entirety of an entity's features, while the "desired composition" encapsulates individual expectations, relevant demands, and societal requirements. From these definitions, we infer that quality is user- or customer-judged, depends on the context, is derived from a multidimensional entity's comparative analysis, and strives to reflect a "degree of excellence" (Engeldrum, 2000; Belmudez, 2015).

When evaluating speech communication networks, one primary performance indicator is the perceived quality of transmitted speech. The ITU-T Recommendation P.10 (2017) defines speech quality as "the quality of spoken language as perceived when acoustically displayed. The result of a perception and assessment process, in which the assessing subject establishes a relationship between the perceived characteristics, i.e., the auditory event, and the desired or expected characteristics."

Essentially, perceived speech quality largely depends on the user's expectations. As the user assesses a speech sample's quality, the physical speech signal initiates the assessment process. The signal is then perceived and interpreted by the listener, resulting in the formation of perceived quality features. Concurrently, a similar process generates the desired quality features based on the listener's experiences. The perceived speech quality ultimately stems from the comparison and judgment between the desired and perceived quality features. As suggested by Wältermann (2013), the perceived quality is a multidimensional value that can be represented as a point in a multidimensional perceptual space.

Research by Wältermann et al. (2010) revealed three orthogonal perceptual dimensions that span the perceptual quality space for speech communication networks: Coloration, Discontinuity, and Noisiness. Côté et al. (2007) identified Loudness as an additional important speech quality feature and proposed it should be included as a dimension. Hence, the perceptual speech quality space spans the following four dimensions:

- **Noisiness** Pertains to degradation such as background, circuit, or coding noise.
- **Discontinuity** Refers to isolated or non-stationary distortions, e.g. due to packet loss or clipping.
- **Coloration** Relates to degradation caused by frequency response distortions, e.g. due to bandwidth limitations, low-bitrate codecs, or packet-loss concealment.
- **Loudness** Denotes non-ideal loudness, either too loud or too quiet signals or loudness variation.

Farias (2010) conducted a thorough examination of video artifacts and provided an extensive overview of the key elements in video quality assessment. Schiffner and Möller (2016), identified four relevant perceptual dimensions through a semantic differential test paradigm followed by a principle component analysis: Incompleteness, Unclearness, Noisiness, and Luminosity. In a subsequent publication, Schiffner employed paired comparison test paradigms with subsequent multidimensional scaling. Taking all his results into account, Schiffner proposes the following set of video quality dimensions for transmitted video to represent the relevant perceptual video quality space (Schiffner and Möller, 2017; ITU-T Recommendation P.918, 2020):

- **Noisiness** Random changes in brightness and color.
- **Discontinuity** Interruptions in the video's flow.
- **Fragmentation** Depicts disjointed, torn or falling apart footage.
- **Unclearness** Describes an unclear or smeared image.
- **Suboptimal luminosity** Represents too high or low brightness.

Addressing the audiovisual quality estimation of interactive multimedia services like video-telephony, Yamagishi and Hayashi (2005) identified two additional dimensions: *aesthetic feeling* and *feeling of activity*. The first dimension pertains to factors like audio and video coding and packet loss, all distorting the audiovisual signal's shape. The second dimension is influenced by factors related to temporal continuity, like one-way transmission delay and video frame rate.

Later, Köster (2017) divides the perceptual quality space of conversational speech quality into three distinct phases that encompass seven perceptual quality dimensions all in all; listening, speaking, and interaction. While the listening phase consists of the four quality dimensions stated earlier for the context of speech communication, the speaking and interaction phases add the following three quality dimensions; see also Köster and Möller (2014):

- Impact of one's own voice on speaking
- Degradation of one's own voice
- Interactivity

2.1.2.2 Signal-Based and Parametric Quality Prediction

The current way of predicting perceived quality is done by signal-based quality estimations or parametric estimators. These estimators, signal-based and parametric, have in common that they analyze technical details and output an MOS that represents an average quality rating. Average here means that, given a reasonably accurate model, the score would be close to what a large crowd of raters would rate. However, no one person is average. Hence, these approaches neglect all the influences human raters experience, all their biases, expectations, and preferences.

This thesis employs two instances from the signal-based category for performance evaluation. As delineated in Chap. 1, the overall objective is to enhance estimation systems that yield a MOS by incorporating individuality data, aiming to increase estimation accuracy for specific individuals (refer to Sect. 2.4 for more details). The effectiveness of this method is tested in Chap. 7 by calculating the discrepancy between the MOS estimations based on signal alone and the new approach presented later, which incorporates individuality information.

For this comparative analysis, two estimators were utilized. The signal-based category was selected as it offers substantial design freedom for experiments. The primary focus was on recording the output signal, independent of the transmission method, streaming software, codecs, or setups employed. Moreover, the two instruments—one for video signals and another for speech signals—were intended to offer state-of-the-art accuracy to underline the significance of improvements derived from the inclusion of individuality data.

An additional criterion was tool accessibility, a practicality concern. In this context, open-source software eased the processing of experimental data, enabling easy testing of various configurations. For these reasons, Non-Intrusive Speech Quality Assessment (NISQA), a signal-based speech quality estimation model, and Disentangled Objective Video Quality Evaluator (DOVER), a signal-based video quality estimation model, were chosen. Nonetheless, the tools' roles are secondary in this thesis, serving merely as references to explore the supplementary value of individuality data in contrast with the standalone performance of these estimation models. The following sections provide a detailed discussion of both tools.

2.1.2.3 Speech Quality Estimation

A commonly used speech quality assessment model was developed by Mittag and Möller (2019). It is non-intrusive and known as NISQA. This innovative model leverages a Convolutional Neural Network (CNN) for estimating per-frame quality and employs a Recurrent Neural Network (RNN) with Mel-frequency cepstral coefficients (MFCCs) input for aggregating these per-frame values over time. Furthermore, the model incorporates a Self-Attention network for time-dependency modeling and Attention-Pooling as a pooling model. The model's design is geared toward regularizing RNN training and minimizing input feature size (Mittag & Möller, 2019).

When compared to contemporary state-of-the-art models of its time, including ANIQUE+ and P.563 (ITU-T Recommendation P.563, 2004), NISQA demonstrated superior performance, achieving an average RMSE improvement of 0.11 (Mittag & Möller, 2019). The model's design caters specifically to monitoring speech quality in samples that lack a clean reference.

The second version of the NISQA project provides estimations for several parameters such as speech quality, noisiness, coloration, discontinuity, and loudness (Mittag et al., 2021). This version was trained and evaluated on 81 data sets and over 13,000 speech files for training and validation.

The NISQA model is available as an open-source library on GitHub for research purposes.[1] Currently, it is under active standardization as work item P.SAMD of ITU-T SG12. The model's code can be used for retraining or fine-tuning, which is a valuable feature, particularly for adapting the model to specific scenarios not initially catered to during its training phase.

2.1.2.4 Video Quality Estimation

For online video sharing and streaming platforms, understanding perceptual video quality, that is, the subjective perception of a user toward video quality, is paramount. This understanding aids in optimizing and enhancing the user experience. Video Quality Assessment (VQA) strives to create a connection between video signals and perceptual quality. It employs objective mathematical models to approximate subjective user opinions (Wang et al., 2021).

DOVER was chosen as an instrumental video quality estimation model due to its performance in the *Video Quality Assessment* challenge presented by *Papers with Code*, a Meta AI Research project, which comprises Machine Learning papers, code, data sets, and evaluation tables.[2] DOVER presents a deep neural network architecture built upon prior works by Wu et al. (2022a). The authors discuss that human quality perceptions are broadly influenced by both technical and aesthetic perspectives in User Generated Content (UGC)-VQA. Hence, they argue, disentangling both dimensions explains humans perceptual mechanisms better (Wu et al., 2022b). Because of this design choice, the quality estimation model reports an aesthetic quality score, a technical quality score, and an overall quality score.

In addition to its estimation accuracy, DOVER was selected for its adaptability. Similarly to NISQA, the source code allows the pre-trained model to be retrained and fine-tuned on specific data sets. This feature proves particularly beneficial in this context, as the model was trained on unspecific, User Generated Content (UGC). Yet, the video-telephony call recordings from the laboratory environment have a specific setup, such as the talking head scenario, consistent lighting, and an uninteresting background. Given that the pre-trained model is not specifically

[1] https://github.com/gabrielmittag/NISQA

[2] https://paperswithcode.com/sota/video-quality-assessment-on-konvid-1k

designed for this type of video, the ability to fine-tune the model on a specific data set makes DOVER a particularly valuable choice. Further details on the fine-tuning process can be found in Sect. 7.1.2.2.

2.1.2.5 Audiovisual Quality and Modality Integration

Detailed analysis and measurement have been conducted on both speech and video modalities, leading to the establishment of fairly precise quality metrics. These metrics are motivated from an understanding of the psychophysical processes involved in auditory and visual perception. In the context of audiovisual perception, however, it remains unclear as to when exactly these modalities are combined during the perceptual processing chain. Consequently, there's a lack of a definitive cognitive foundation that precisely explains how users of multimedia services amalgamate information from diverse sources such as audio, video, and haptic, to form an overall quality judgment (Reiter & You, 2010).

In a bid to understand the audiovisual perception process, researchers have gravitated toward attention theories. The widely accepted theory, known as late fusion, posits that auditory and visual signals undergo separate internal processing to generate distinct auditory and visual qualities. These are then fused at a late stage to render a judgment on the overall perceived quality (You et al., 2010). Therefore, audiovisual quality is typically depicted as an amalgamation of two dimensions: speech and video qualities (Belmudez & Möller, 2013).

Moreover, the evaluation of human perception of audiovisual quality is reliant on the experimental methodology employed. Any subjective tests conducted should adequately represent the ecological environment of the service or application being assessed. Research by Belmudez and Möller (2013) presents how perceived auditory and visual qualities are integrated to form an overall perception of audiovisual quality in varying experimental contexts. The research proposes a universal integration function for two types of experiments: passive viewing/listening and interactive contexts.

The integration function is represented by the following equation for an interactive scenario:

$$MOS_{AV} = \alpha MOS_S + \beta MOS_V + \gamma MOS_S * MOS_V + \zeta \tag{2.1}$$

Here, MOS_{AV} signifies the audiovisual quality, MOS_S represents the speech quality, MOS_V is the video quality, and α, β, γ, and ζ are the scalar coefficients. Several preliminary experiments were performed in the 1990s to derive this mathematical formula that carries out audiovisual integration based on the quality metrics of the individual modalities (ITU-T Contribution COM 12-64-E, 1997, 1998; ITU-T Contribution COM 12-61-E, 1998). These experiment results led the International Telecommunication Union (ITU) to suggest using only the multiplicative term between the speech and video qualities ($MOS_S * MOS_V$) with an additive shift as an estimate for the audiovisual quality (ITU-T Recommendation P.911, 1998).

In the study by Belmudez and Möller (2013), the authors determined the coefficients for an interactive, conversational setting. The resulting integration function is as follows:

$$MOS_{AV} = 0.030 * MOS_S + 0.079 * MOS_V + 0.123 * MOS_S * MOS_V + 1.374 \quad (2.2)$$

Though integrating speech and video quality estimations remains the standard practice, there are attempts to derive audiovisual quality estimation using a single mode. One such instance is the NAViDAd system by Martinez et al. (2019). It calculates a no-reference audiovisual quality metric leveraging a deep autoencoder. The video component's visual features are determined by natural scene statistics and spatial-temporal measures, while the audio features are obtained by calculating the audio component's spectrogram representation. The autoencoder layer produces a compact set of features, subsequently mapped to audiovisual quality scores by the classification layer. The results suggest that this approach effectively captures the intricate interplay between audio and video components.

2.1.2.6 Subjective Influences on Experience Perception

The perception of QoE is influenced by a myriad of subjective factors that vary from personal attributes like age, gender, and cultural background, to psychological states such as mood and expectations. According to the Qualinet White Paper on Definitions of Quality of Experience, these factors are regarded as any characteristic that may influence the QoE for the user (Qualinet, 2013).

The ITU-T lists the "characteristics of the application or service, context of use, the user's expectations with respect to the application or service and their fulfillment, the user's cultural background, socio-economic issues, psychological profiles, emotional state of the user, and others" as influencing factors on QoE in their vocabulary definition recommendation (ITU-T Recommendation P.10, 2017).

In *Factors Influencing Quality of Experience*, Reiter et al. (2014) categorize these Influencing Factors (IFs) into three groups: Human, System, and Context. They argue that these three categories collectively form an experience. This book primarily focuses on the "human" dimension of this triad. While the system's majority aspects can be objectively analyzed (e.g., the audio and video stream), the context and human aspects necessitate careful observation and description to accurately estimate a QoE rating at an individual level (Jumisko-Pyykkö, 2011).

There is ample literature illustrating the correlation between human aspects and QoE perception. For instance, personality traits can predict behaviors, decisions, and reactions in VR environments (Aranha et al., 2018), and influence the level of presence experienced (Galloso et al., 2016). The findings of Schmitt et al. (2017) further support this by showing that user engagement and enjoyment during videoconferencing are influenced by individual differences, thereby affecting QoE. Their predictive models incorporating user background and current state have shown significant improvements in accuracy, underscoring the importance of these factors in QoE assessment. User characteristics like agreeableness, technology acceptance, and perceived ease of task completion can impact positive user experiences with intelligent robotics (Bernotat & Eyssel, 2017). Subjective influences, like the "halo effect," can extend a positive impression of a service or product to more specific judgments about that entity, influencing the QoE (Kahneman, 2011). Expectations, guided by the "expectation-disconfirmation paradigm," play a crucial role in shaping the perception and QoE (Oliver, 1980). The user's emotional state also significantly influences QoE perception. A positive mood may enhance the perception of quality, regardless of the stimulus, and vice versa (Russell, 2003). Demographics such as age and cultural background can also influence QoE perception. Younger users, for instance, may have higher expectations toward technical services, influencing their QoE perception (Morris & Venkatesh, 2000; Venkatesh et al., 2012). Cultural differences can potentially impact QoE due to varying expectations and interpretations of quality (Hofstede, 1984). Perception is also influenced by past experiences and education, which are seldom identical among individuals (Démuth, 2013). Additionally, timing significantly alters the perception of experiences. The phenomenon known as the *recency effect* demonstrates how recent ratings can significantly influence subsequent ratings (Hands & Avons, 2001; Bampis et al., 2017).

This extensive body of literature signifies the profound role individual differences play in QoE perception. Importantly, QoE is not a rigid, universally perceived construct; it is highly subjective and individual. Therefore, accounting for such individuality while estimating QoE is crucial to narrowing the gap between estimated and actual user ratings.

2.1.2.7 Physiology and Behavior-Based Quality Prediction

A 2016 study proposes the use of affordable consumer electronics to capture objective physiological metrics, Heart Rate (HR), and Electrodermal activity (EDA), as users consume VR content. Results show that HR and EDA measures distinguish the VR and non-VR environments and that EDA was more representative of user QoE than heart rate, with higher QoE ratings for users of an HMD compared to a 2D display screen (Egan et al., 2016). While the subject was linked to VR environments, it demonstrates how EDA and HR measures can be used to measure a person's QoE. However, the use of HR as a physiological metric to infer experience alone is problematic as it is influenced heavily by individual differences such as how regularly an individual engages in sports. Schmidt et al. (2020) looked at Heart Rate Variability (HRV) as a metric in a QoE scenario. They offer evidence that network impairments can be crucial for immersive experience, and continuous HRV measures can supplement self-reported measures to accurately assess the immersive experience of cloud gaming. While their application is also not video-telephony, they show how Electrocardiogram (ECG)-based physiological monitoring can be employed to objectively measure a person's experience.

Arnau-González et al. (2017) proposed using physiological signals, such as ECG, Electroencephalography (EEG), and Electromyography (EMG), to develop a machine learning model that can map features derived from these signals to a subjective video quality scale. Using the physiological recordings, they classified the perceived quality of a video sequence as low or high compared to the original reference video sequence. Results indicated that a combination of the features extracted yielded the best results and point toward a feasibility of the physiological dimensions used. However, this pilot study only included 12 participants, and they tested a binary classification instead of an estimation of continuously scaled QoE ratings.

The work by Donley and colleagues (2014) investigates the QoE of multi-sensory media through analyzing biosignals collected by EEG and eye gaze tracking and comparing them with subjective ratings (Donley et al., 2014). The impact on QoE of various levels of synchronicity between added sensory effects (wind, vibration, and light) and target video scene was also investigated. While there was no statistical difference observed for the QoE ratings for different levels of sensory effect synchronicity, an analysis of raw EEG data showed differences in physiological responses. The eye gaze data showed more deviation for a video with synchronous effects, and the EEG showed correlating occipital lobe activity for this instance. These differences in physiological responses indicate that sensory effect synchronicity may affect QoE despite subjective ratings appearing similar (Donley et al., 2014). This makes a strong case for the effectiveness of using biosensors in evaluating viewers' subjective experiences—although the study focused on binary classification, instead of utilizing the physiology sensors to regress actual QoE ratings.

Arndt et al. (2012) provided convincing proof of the efficacy of employing Electroencephalography (EEG) as a physiological means to assess quality perception. Their results indicated a positive correlation between the level of video degradation and the early rise and amplification of P300 amplitude, an event-related potential that manifests as a positive waveform around 300 milliseconds post-stimulus. Furthermore, they found a high correlation between peak amplitudes and the Mean Opinion Score (MOS) for audiovisual content. Similarly, Scholler et al. (2012) reliably detected variations in the quality of brief video stimuli through an analogous EEG analysis.

In a later study, Arndt et al. (2014b) leveraged spatially degraded videos to gather subjective quality judgements, eye movement parameters, and brain activity data via EEG. They suggested a model that predicts quality scores from physiological measures, demonstrating that such measures serve as a compelling addition to standard methods, as they reveal information otherwise unobtainable through subjective approaches. Their linear regression model accounted for 41.4% of variance in predicting the Degradation Category Rating (DCR) score, using alpha proportion (the relative quantity of alpha brainwave activity present in an EEG signal) and pupil dilation as indicators. Although this research comes close to using physiological measures for estimating quality ratings, it deviates considerably from assessing the QoE scores of distorted audiovisual content. This is because the researchers introduced a visual search task by segmenting the visual field into nine equidistant areas, distorting only one segment, and requiring participants to locate and rate the distorted area. Despite these differences, this research demonstrates how pupillometry and EEG could be utilized to monitor QoE impressions via physiological signals, see also Antons et al. (2014).

2.2 Psychology, Psychophysiology, and Behavior

Section 2.1.2.6 briefly touched on the diverse influences or biases that alter an individual's perception of a signal's quality. This suggests that humans don't perceive the same signal equally, underscoring the importance of understanding the fundamental concepts of how individual differences influence perception. In this section, I delve into pertinent concepts related to individuality in the fields of psychology, psychophysiology, and behavior.

Psychophysiology, a field rooted in psychology, investigates the complex interplay between psychological processes and their physiological underpinnings. It studies these interactions within the confines of an intact organism (Furedy, 1983). A critical component of psychophysiology is evaluative reporting, characterized by introspective participant responses and self-ratings of internal psychological states or physiological sensations. For instance, self-reporting of arousal levels through questionnaires.

Individuality can be categorized using different types of personal data that encapsulate unique physiological and psychological traits and states. These data help describe who we are and how we feel. Concurrently, responses to stimuli can be registered as physiological (subconscious) and behavioral (both subconscious and conscious) reactions. These data sources offer a holistic view of an individual, irrespective of specific situations or stimuli.

Despite these insights, the existing literature falls short in providing clear-cut guidelines for the dimensions or measurement tools for these data categories. The predictive value of these data classes for endogenous variables like audio-visual quality or valence and arousal ratings is acknowledged, but they do not explain a significant portion of the variance. As I discuss in Sect. 5.1.2.2, trait and state variables alone predict audio-visual quality ratings, valence, arousal, listening effort, and mental load with adjusted R^2 values ranging from 0.03 to 0.19 in a video-telephony, short conversation setting.

2.2.1 *Traits and States*

Traits are generally stable personality characteristics that are unique to an individual. These can include traditional personality dimensions such as openness, agreeableness, or neuroticism, as well as other aspects such as intelligence or chronotypes.

States, on the other hand, are short-term phenomena like emotions, hunger, or sleepiness. These states can fluctuate rapidly, from changes within seconds (like the affective response to a spontaneous stimulus) to changes within a day (like sleepiness).

The modest predictive power of traits and states is understandable considering their distinctive, stimulus-independent nature. They don't act as standalone predictors. Instead, they modulate continuously observable physiological and behavioral responses.

2.2.1.1 Traits

Personality traits characterize individual uniqueness and are generally consistent over time. The Five-Factor Model, also known as the Big Five, is a frequently cited framework for classifying personality traits, which include openness, conscientiousness, extraversion, agreeableness, and neuroticism (Costa Jr & McCrae, 1992). Additional personality elements, such as susceptibility to frustration and intelligence, also qualify as traits. These elements tend to remain stable for years or even a lifetime, rather than altering over brief periods.

The role of traits in shaping quality perception has been demonstrated by research, for instance, Scott et al. (2016) found that personality and cultural traits explain 9.3% of the variation in quality ratings. Models that integrate human factors for predicting perceived enjoyment exhibit 3%–9% enhanced performance compared to those solely reliant on signal information. It is noteworthy, however, that the Six Factor Model of cultural differences, Hofstede (2011), is not intended for individual usage, and application in such a context is generally dissuaded.

2.2.1.2 States

The subject of affective states, comprising of emotions, moods, and related experiences such as feelings, covers both physiological responses and cognitive processes. The physiological responses include alterations in heart rate or hormone levels, while cognitive processes entail the appraisal and interpretation of events.

Within this work, emotions are regarded as transient experiences, their duration ranging from mere seconds to several minutes. Typically, they are acute reactions to specific events or objects, and their focus remains fixated on these stimuli (Reiter et al., 2014). Moods, however, stand in contrast. Their existence is more enduring, persisting for hours to several days, and they are not incited nor directed toward a singular, specific object or event (Frijda, 1994).

One pivotal component in studying emotions is affect, traditionally described as having two major components: valence and arousal. The former, valence, refers to the positive or negative quality of an emotion, while arousal speaks to the degree of wakefulness or activation accompanying an emotion (Russell, 1980). Russell's circumplex model of affect, along with Russell's affect grid (Russell et al., 1989), is routinely employed to assess emotional responses following distinct experiences.

Valence and arousal have demonstrated enduring correlations with quality ratings, with valence presenting particularly powerful associations. As substantiated by Gupta et al. (2015) and later presented results of this work, these elements play a critical role in estimating an individual's perceived QoE.

Overall, states, unlike traits, are subject to change and do not remain a fixed attribute of an individual. Moreover, they can be influenced by current circumstances or conditions, although this influence is not always necessary.

2.2.2 *Individuality*

Individuality refers to the qualities and characteristics that differentiate one person from another. It is a multifaceted concept, influenced by a range of factors including:

- **Personality Traits:** Stable characteristics that influence behavior across a variety of situations (Costa Jr & McCrae, 1992).
- **Beliefs and Values:** Including religious and spiritual beliefs, moral values, political ideologies, and beliefs about self and the world (Schwartz, 1992).
- **Interests and Preferences:** Such as hobbies, foods, music, movies, etc. (Holland, 1997).
- **Skills and Abilities:** Including intellectual, artistic, athletic, or other skills and abilities (Carroll et al., 1993).
- **Life Experiences:** Including upbringing, education, relationships, job experiences, travel experiences, and more (McAdams, 2001).
- **Biological Factors:** Including genetics, brain structure, and physical characteristics (Plomin & Daniels, 1987).
- **Cultural and Social Influences:** Including cultural background and social environment (Markus & Kitayama, 1991).

The broad range of influences contributes to the extensive array of individual differences, which form an intrinsic part of human nature. These influences serve as the foundation for selecting relevant dimensions for the studies explored in the upcoming chapters. Although I will only draw upon a select few from this range, referring to this list may prove beneficial in other experimental contexts.

2.2.3 *Individual Differences and Theoretical Models*

Understanding individual differences is paramount in comprehending how a single situation, stimulus, or environment elicits diverse experiences among individuals. For instance, Döring et al. (2022) proposed a four-dimensional model considering both technological and personal factors, discerning between individual and societal influences while exploring fatigue in videoconferences. This model builds on the Differential Susceptibility of Media Effects Model (DSMM), which delves into individual responses to media stimuli and how social context impacts these experiences (Valkenburg & Peter, 2013).

One proposition asserts, "Media effects are conditional; they depend on three types of differential-susceptibility variables: dispositional, developmental, and social" (Valkenburg & Peter, 2013). Here, media effects represent changes, short-term or long-term, within individuals due to media exposure. Dispositional susceptibility encompasses personal characteristics such as personality traits and states, like moods. Developmental susceptibility arises from an individual's cognitive, emotional, and social development throughout their lifespan. Social susceptibility draws from social context factors, which include interpersonal relationships, institutional influences, and societal norms.

Another proposition posits that "Media effects are indirect; three media response states (cognitive, emotional, and excitative) mediate the relationship between media use and media effects." The cognitive response state includes attention, comprehension, and cognitive engagement with the media content. It illustrates how individuals selectively attend to and process elements like the message, storyline, character motivations, and perspectives. The emotional response state captures the emotional reactions elicited by media consumption, such as feelings of joy, sadness, fear, or anger. Lastly, the excitative response state embodies physiological arousal and excitation, potentially influenced by media content that is stimulating, arousing, or intense.

Per the DSMM, these three response states mediate the relationship between media use and media effects, significantly shaping individual cognitive, emotional, attitudinal, and behavioral changes resulting from media engagement. By accounting for these response states, the DSMM offers a framework for understanding the variation in media influence on individuals. It underscores the critical role of individual characteristics and social contexts in determining susceptibility to media effects and the ensuing cognitive, emotional, and excitative responses mediating these effects (Valkenburg & Peter, 2013).

Although the DSMM doesn't directly address the effect of individual differences in QoE formation, it correlates individuality with diverse behaviors and reactions in media use. This is particularly pertinent for this thesis, as the model also implies a connection between individuality, media use, and response states. The relationship between these elements and QoE ratings remains unclear, but this model serves as a crucial theoretical basis to integrate individuality and potential response states.

Other theoretical models, such as the *Unified Theory of Acceptance and Use of Technology* by Venkatesh et al. (2003) and the *Attention Restoration Theory* by Kaplan (1995), emphasize the importance of individual differences in perceiving and experiencing media. Most prominently, sociodemographic variables, personality types, and cognitive traits are identified as significant factors that influence the perceptual aspects of a user's experience.

2.2.3.1 Influence of Individual Differences on Experiences

In the context of audio-visual experiences, speech quality, intelligibility, and listening effort have shown to be interrelated (Krueger et al., 2017). Low-quality speech and audiovisual stimuli can lead to fatigue (Antons et al., 2012a; Arndt et al., 2014a, 2016).

Current objective, signal-based estimators cannot account for these individual differences. They are designed to estimate a MOS. When researching how physiological analyses can improve quality estimation, it is important to remember that each user is unique and will produce different sensor readings when presented with the same stimuli. These individual responses need to be taken into account for a more accurate assessment of the user's experience.

2.2.4 *Physiological and Behavioral Responses*

The human body is made up of highly individual, complex mechanisms in which it is difficult to identify discrete states without a battery of tests. However, the body does show multiple indications of its states, known as physiological signals, produced by the body's physiological processes (Airij et al., 2020). For example, the limited capacity model of mediated message processing proposes that multimedia experiences influence cognition, emotion, and physiological arousal, which subsequently affects attention and memory processing (Lang, 2000).

For example, correlates have been found between psychophysiological responses Valence, Arousal, Flow, Fatigue (Punsawad et al., 2011; Antons et al., 2012a). The purpose of recording physiological and behavioral responses is to provide continuous information about a user's cognition. Some responses are directly linked to cognitive functioning, such as EEG, while others are merely proxies of cognitive processes, such as EDA. While professionals in the QoE domain are mostly interested in the experience of a situation or a stimulus, finding bodily correlates to these experiences allows for estimating impressions without interrupting the consumption of a stimulus or a call situation. This also allows to estimate impressions continuously over time. Moreover, such correlations would allow for more accuracy and individualized estimations rather than predicting an MOS.

Engelke et al. (2016) suggest that the concurrent use of both methodologies might lead to deeper insights into high-level QoE attributes. They argue that while psychophysical assessment has been useful in evaluating user experience, it may not be enough when it comes to the evaluation of advanced media technologies. Traditional methods such as psychometric scales and surveys depend on conscious responses and often do not provide sufficient insight into underlying perceptual and cognitive processes. This issue is particularly true for high-level QoE concepts that are difficult to quantify using traditional psychophysical methods. To overcome this challenge, the chapter suggests that other implicit methods, such as those based on psychophysiology, are needed for QoE assessment. Psychophysiological methods measure implicit rather than explicit responses to physical stimuli. These methods can provide insights into underlying perceptual and cognitive processes that would be difficult to capture using traditional psychophysical methods and subjective questionnaires. The use of these methods in combination with the standard psychophysical methods could, therefore, lead to more profound insights into high-level QoE attributes in advanced multimedia technologies (Engelke et al., 2016).

The list of potential physiological dimensions is long and also not clearly defined. However, some make the distinction between physiological and behavioral responses. In the context of this work, consciously influenceable dimensions such as gaze or breathing frequency are behavioral. In contrast, those that are not consciously changeable, such as the diameter of the pupil, heartbeats, or brain waves, are physiological signals. Following, methods used are discussed in detail. However, there are numerous additional measures that were not considered in this work. The rationale for selecting one over the other is primarily practicability-driven: While focusing on the most popular dimensions used in the literature, there are specific challenges in conversation experiments. These experiments typically take long to carry out, as two participants have to arrive on time, have to be introduced to the experiment, have to be wired-up with the sensors, and so on. Hence, measurements that do not involve physical contact are preferable, as they tend to be easier to handle and set up. This rules out EEG since it is a complex and time-consuming procedure to equip the participants with. Another consideration is the naturalness of the appearance of the participants. To resemble a call situation as naturally as possible, I tried to not obstruct the participant's faces with unnecessary technical equipment, other than headphones. This also meant not to use any head-mounted eye-trackers, but only stationary ones. These considerations lead me to the following list of measurements.

2.2.4.1 Electrocardiogram: ECG

ECG is a very popular physiological dimension, as it is easy to record and well studied. Apart from more complex clinical 12-lead ECGs, there are several studies demonstrating good accuracy even for simpler, consumer-grade wearables like chest straps or even Photoplethysmographie (PPG)-based wrist-worn smart watches (Montalvo et al., 2022). Especially for simple beat-to-beat analyses (neglecting other information provided by the PQRST-complex of a clinical ECG) and for low-movement scenarios, chest straps such as the Polar H-10 can surrogate a clinical ECG (Montalvo et al., 2022). The beat-to-beat analyses are especially interesting to derive HRV metrics. These quantify changes in heartbeats, and there is a body of literature discussing the predictive value for cognitive states like cognitive load, stress, or even affective states of HRV analyses. For example, Bolls et al. (2001) used, among others, heart rate data to test if radio advertisements elicit physiological reactions in listeners and suggest that negative messages receive more attention than positive messages, based on their heart rate analyses, and in Spang et al. (2022a) we discussed the link between stress and HRV.

For the experiments discussed in the upcoming chapters, a Polar H-10 chest strap was used. This provides a single-lead ECG, samples internally with 125 Hz, and provides an internal peak detection to report RR-Intervals (RRI), which are used to base the HRV analyses on. The device is very easy to apply; it is mounted to a flexible band that participants wear around their chest on their skin. The band can be disinfected and cleaned effectively and quickly and the device provides a reliable signal. This signal was streamed via Bluetooth to a separate recording device.

2.2.4.2 Electrodermal Activity: EDA

EDA is a commonly used psychophysiological measure reflecting changes in the electrical properties of the skin that are closely tied to the activity of the Sympathetic Nervous System (SNS). This makes EDA a crucial tool for indirectly assessing changes in physiological arousal linked to a wide range of cognitive and emotional states (Boucsein, 2012). EDA is often referred to by a variety of other terms including galvanic skin response (GSR), skin conductance (SC), or psychogalvanic reflex (PGR).

2.2.4.3 Eye Tracking and Pupillometry

Eye tracking, including gaze tracking and pupillometry, is a powerful, noninvasive method for examining cognitive processes and states. This physiological measure assesses where an individual looks (point of gaze), the movement of an eye relative to the head, and changes in pupil size, which is known as giving insight into cognitive load and effort (Duchowski et al., 2018).

Underpinning the science of eye tracking are various eye phenomena such as saccades (rapid eye movements), fixations (periods of relative eye stability that facilitate information intake), and blinks, which are thought to reflect cognitive processes, particularly cognitive load. Among these measures, pupillometry—the measurement of pupil diameter—has gained popularity due to its sensitivity to changes in cognitive load. The pupils dilate and contract under the control of the autonomic nervous system, with cognitive effort, emotional state, and light conditions among the influencing factors.

Hess and Polt (1964) were among the first to demonstrate the relationship between pupil dilation and problem difficulty. Their findings were corroborated and expanded by Kahneman and Beatty (1966), who suggested that pupil diameter provides a "very effective index of the momentary load on a subject as they perform a mental task." Ahern and Beatty (1979) and Beatty (1982) further refined this metric, referring to it as Task-Evoked Pupillary Response (TEPR), emphasizing its potential as an indicator of cognitive load. This relationship is believed to reflect central, or brain-based processes, rather than peripheral ones (Beatty & Lucero-

Wagoner, 2000). In the context of Cognitive Load Theory (CLT) by Sweller (1988) and Chandler and Sweller (1991), which focuses on estimating human workload in learning tasks, eye tracking, particularly pupillometry, has shown promise as a noninvasive measure of cognitive load (Duchowski et al., 2018).

Eye tracking and pupillometry could measure cognitive load during video calls and potentially elucidate individual differences in cognitive responses to quality degradation in video-telephony calls. The interpretation of such data, in tandem with other physiological and behavioral measures, could provide a comprehensive understanding of user experience during video-telephony calls.

2.2.4.4 Face Tracking

Face tracking is a burgeoning technology that utilizes computer vision and machine learning to identify and track facial features, thereby offering insights into various cognitive states including emotional responses, mental load, effort, and attention.

Traditionally, facial expressions have been a principal focus in this field. These expressions, such as frowning or gaze aversion, have been used to infer mental effort and cognitive load, as well as emotional responses in various contexts (D'mello & Graesser, 2010; Lewinski, 2015; Girard et al., 2015).

The interpretation of facial expressions has been a subject of ongoing debate in the field of neuroscience and cognitive psychology. Researchers like Feldman-Barett propose the term *facial configurations* in lieu of "facial expressions." They posit that the presumed one-to-one mapping between specific facial expressions and distinct emotional states may not be universally applicable. Instead, the term *facial configurations* describes patterns of facial muscle movements that do not necessarily imply a specific emotion, thus decoupling the facial appearance from any presumed emotional state (Barrett, 2017; Barrett et al., 2019).

However, the capacity of face tracking to measure viewer engagement through facial motion was demonstrated by Abadi et al. (2013), who used a real-time facial expression recognition system that tracked 12 facial interest points. The motion of these points was used to derive features encoding viewers' engagement levels.

It is crucial to consider the limitations of relying solely on face tracking for gauging emotional states and engagement. Some studies have noted the difficulty of reliably detecting certain affective states, such as boredom and engagement, from facial cues alone (Craig et al., 2004, 2008; McDaniel et al., 2007). Furthermore, the lack of clear facial expressions does not necessarily equate to the absence of emotional states (De Moor et al., 2014). Therefore, it is important to combine face tracking with other measures such as self-reports, eye tracking, or physiological measures, to create a more comprehensive profile of a participant's experience.

In the context of this work on quality degradation in video-telephony calls, face tracking can offer valuable insights into participants' reactions and perceived cognitive load in response to various video quality scenarios. Like eye tracking, face tracking employs non-intrusive, camera-based technology, obviating the need for physically attaching sensors or electrodes to participants (Duchowski et al.,

2018). Machine learning models process the camera images to provide real-time information on various facial features or *configurations*, such as the degree to which an eye is open or the intensity of a smile. This provides a wealth of detailed, user-friendly, and accessible data about participants' experiences without imposing physical discomfort.

2.2.4.5 Conversational Interactivity

Given the focus of this manuscript on video-telephony scenarios, and therefore the need to quantify various aspects of behavior in a conversational scenario, the dynamics of conversation become critical aspects to consider, ideally serving as a window into participants' experiential dimensions.

Kiousis (2002) defines many different types of interactivity used in the literature, although none of them seem viable for the purpose of describing interactivity in the telephony-based conversational scenario. Thus, Hammer (2006) defined conversational interactivity as "*a single scalar measure based on the quantitative attributes of the participants' spoken contributions*."

Hammer and Reichl (2005) have pioneered the development of conversational interactivity metrics, particularly exploring their impact on the perceived Quality-of-Service in bi- and multilateral communication over delayed transmission channels, such as telephony. They modeled the dynamism and fluidity present in a two-way conversation using a stochastic process with four states: person A speaking, person B speaking, mutual silence, and double talk. Conversational interactivity, in their perspective, is characterized by three key properties: the basic relationship between sojourn times (the time a system spends in a particular state) and interactivity, the rate of information flow, which reflects the aspect of speed, and the transitions between the different states and their probability of occurrence, encapsulating the essence of interactivity (Hammer & Reichl, 2005).

Building on this, they introduced a novel metric—the "Conversational Temperature"—designed to measure the level of interactivity during telephone conversations. The metric, based on sojourn times as recommended by ITU-T, is intended to scale with a norm conversation, thus providing a standard benchmark. The first-order behavior shows that increased sojourn time in one state leads to decreased interactivity and vice versa, underscoring the fluid and dynamic nature of conversation (Hammer et al., 2005). The Conversational Temperature Metric is further discussed in Reichl and Hammer (2004).

The concept and metrics of conversational interactivity provide a rich set of tools for analyzing communication dynamics in video-telephony and can be especially insightful when studying the impacts of quality degradation. By capturing the fluid and interactive nature of the conversation, these metrics can offer nuanced insights into how video-telephony quality influences the conversation dynamics and, ultimately, the perceived Quality of Experience.

2.3 Other Terms and Definitions

The concepts of **estimation** and **prediction** form the crux of making forecasts based on available information. The objective of estimation is to determine unknown parameters or infer characteristics using accessible data and models, typically used in statistical analysis to infer population parameters. Prediction, however, revolves around projecting future outcomes using existing data and models, instrumental in forecasting human or system behavior and forming hypotheses about future events based on observed data patterns.

In the scope of this thesis, estimation takes precedence. It analyzes systems, identifies unknown parameters, and infers characteristics from accessible data. In the realm of signal-based methods, a human auditory system model is applied to the degraded received speech or video signal. Then, the perceived quality is *estimated* using a similarity measure calculated from a psychoacoustic representation of the received speech signal (Hammer, 2006).

Conversely, prediction concentrates on making informed decisions and projecting future events. For instance, parameter-based methods *predict* perceived quality based on the transmission system's characteristics (Hammer, 2006).

The terms **subjective** and **objective** are also crucial in this context, each having two dominant interpretations. Initially, *subjective* is associated with anything human-derived. A rating scale, much like a measured HR, is a *subjective* measurement, even if recorded by a machine. Any measurement involving human participants would be deemed *subjective*. In contrast, *objective* refers to a reproducible study setting, ensuring reliable measurements and reproducibility of results. Distinctions can be made between *voluntary* or *volitionally* influenced measurements and *involuntary* or *automatic* observations, but these do not precisely distinguish variable types, acquisition methods, or interpretations.

Another interpretation of these terms defines *subjective* as anything consciously processed, interpreted, and potentially influenced by a participant, such as questionnaire responses. Conversely, *objective* measures are involuntary, uncontrolled, and typically recorded by a device, like HR analysis. Following this definition, the reproducibility of the experiment does not feature in either category, but it does differentiate conscious participant interactions from involuntary or automatic responses. I adopt this latter interpretation throughout this thesis, where *subjective* measurements are those participants consciously control, and *objective* measures are involuntary records, typically device-facilitated.

In the context of algorithmic quality estimation methods, the outputs are *objective*. These parameter- or signal-based tools may be trained on subjective data to replicate human ratings, but they exclude the human rater who consciously experiences a stimulus. Such quality estimation methods are known as **instrumental** methods, following Raake (2006).

2.4 Objective and Research Questions

Throughout the literature, evidence suggests that individuals' experiences and perceptions are subject to numerous influences. However, standardized approaches to QoE assessment tend to overlook these complexities. Notably, while many smaller studies investigate these influences, there is a lack of comprehensive research considering a larger number of variables simultaneously. Despite this, some dominant and recurring factors consistently demonstrate a correlation with individuals' QoE ratings. It is thus evident that these influences significantly shape a person's individual QoE rating for a given situation or multimedia artifact.

While existing literature provides complex conceptual models describing the formation of QoE ratings, they tend to neglect the integral aspect of human individuality. When considered at all, individuality is treated as a vaguely defined influencing factor. However, given its clear relevance, my primary research question arises:

1. Research Question—the model
How would a QoE formation process model look like, which focuses on individuality influences?

A model responding to this question requires empirical validation and needs to prove its soundness statistically. It is not intended to replace existing comprehensive models, but to enhance them by focusing on aspects of individuality and their assessment. This would ultimately enable a tool useful for QoE assessment projects. As my goal is to develop and assess such a model, a second research question presents itself:

2. Research Question—the influences
Which individuality factors have the most dominant influence on QoE ratings?

The practical relevance of the model hinges on its ability to enhance QoE rating estimations over existing procedures that do not consider individuality. Therefore, the following research question relates to the estimation accuracy of the model:

3. Research Question—the performance
How relevant is an individuality-informed QoE estimation, as proposed by the developed model?

As part of empirically substantiating a model tailored toward individual influences on QoE, it becomes crucial to identify which influences affect the QoE rating most. The answer to this question will be sought considering a wide array of variables, the detailed discussion of which is reserved for the upcoming Chap. 4. The empirical validation, however, is not confined to a single experimental context. Two paradigms are to be explored: a video-based telephony simulation for controlled laboratory conditions and a real-life conversational scenario to authenticate the simulation findings. This leads me to an additional research question:

4. Research Question—Viability of simulations
Can a video-based conversation simulation be considered a reliable method to study human reactions to quality degradation?

Chapter 3
The Individuality Assisted Estimation Model of Experience

The words printed here are concepts. You must go through the experiences.
Saint Augustine

This chapter presents a hypothetical, structural model, called IAE model, that explains how an audiovisual stimulus and individual factors contribute to the formation of an experience rating. In addition to former models explaining how an experience rating is formed, the IAE model also takes into account individual factors like individual traits, and experience dependent and independent temporal states, which are not assessed in any other QoE model. The model's assumptions are based on current research in the field. It unifies components that have been thought of in the past, but, to the best of my knowledge, have never been combined to focus on individualized QoE estimation. I hypothesize, that individual aspects, observable responses, and the stimuli themselves together enhance current quality estimation models, a hypothesis that is thoroughly tested in Chap. 7. In this chapter, the hypothetical model and its theoretical foundation are laid out.

3.1 Motivation

Previous models of experience, such as those discussed in Sect. 2.1.1, focused on the experiences themselves rather than how they can be observed through proxies like physiological responses or individual influences. This chapter aims to explore the possibilities of QoE measurements by incorporating physiological and behavioral responses into a model that describes the formation of reactions and their influences on the human body. Existing research lacks models that predict or estimate QoE dimensions for individuals. Instead, they looks at humans as average humans, whose experience is only influenced by factors being based in the

R. P. Spang, *Individualized Quality of Experience Estimation in Audiovisual Communication*, T-Labs Series in Telecommunication Services,
https://doi.org/10.1007/978-3-031-67551-5_3

stimulus or its transmission, e.g., Raake and Egger (2014). However, a shift toward individualization of experience metrics can improve service customization and meet individual needs. This argument is relevant both from a business perspective and an academic standpoint. Furthermore, understanding individual differences can help differentiate between the impact of stimuli and personal characteristics on ratings or responses.

For instance, individuals with positive personalities may consistently perceive the same stimuli or conditions more positively than those with a negative outlook. Previous studies have shown that human factors account for a significant portion of the variance in quality ratings. As discussed in Chap. 2, there's a body of literature suggesting a direct influence of human factors on perceived multimedia quality. Hence, it is fair to hypothesize that these interpersonal differences influence rating behavior, although the stimulus remains the same. It is the persons' influence that makes them have a different experience. For instance, Scott et al. (2016) found personality and cultural traits to account for 9.3% of the variance of quality ratings. Models predicting perceived enjoyment, given media-quality, performed between 3% and 9% better when human factors were included, compared to purely signal-based models. Therefore, there is a need to focus on individual differences and explore how individualization of quality estimation can be achieved and implemented.

Therefore, as stated in Sect. 2.4, the goal of this work is to develop a model that explains individuality influences and provides a framework for understanding rater differences with identical stimuli. Additionally, the model aims to promote the recognition of individual differences and their utilization in making products, services, and scientific studies more accurate. The hypothesis is that a model considering individual differences will improve the accuracy of estimating true user ratings compared to a one-size-fits-all model. In other words, incorporating information about the person experiencing a stimulus or situation can generate results closer to their actual rating.

3.2 Development

To define terms and concepts, I break down the complex process of forming an experience rating from exposure to a stimulus into fundamental steps. This pathway model follows a linear trajectory, although there can be overlaps. Figure 3.1 visualizes the key processes involved.

As Privitera (2018) put it: "Amazingly, our senses have the ability to convert real-world information into electrical information that can be processed by the brain. The way we interpret this information—our perceptions—is what leads to our experiences of the world." The first sentence, here, describes *sensation*. This then is *interpreted* to become a *perception*—if *attention* is being paid.

Here is an overview of the most important processes involved:

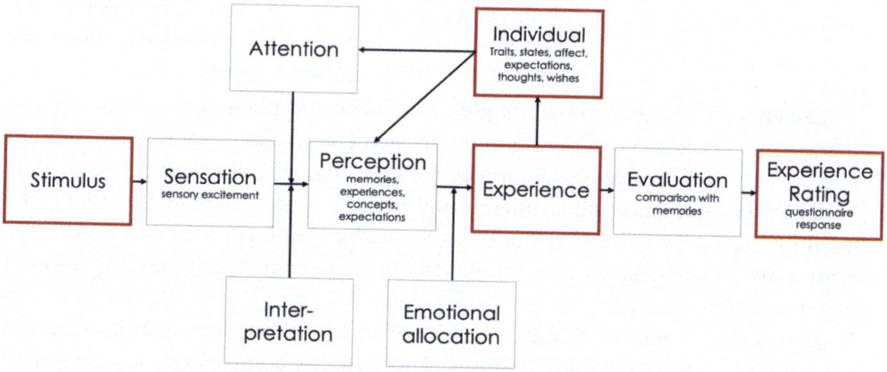

Fig. 3.1 Pathway model describing how a stimulus forms an experience and how a user rating of said experience is formed. The model describes the elements involved, starting with a stimulus evoked sensory excitement, over perception to experience. This experience is then compared against previous experiences of similar situations to derive a rating

1. **Stimulus:** A stimulus is presented. Sensory organs receive stimuli from the environment, such as visual or auditory signals.
2. **Sensation:** This is the first step in the actual process. Sensation refers to the physiological response to a stimulus. For example, when someone touches a hot object, their skin senses the heat; when someone sees a bright color, their eyes capture the light. This process involves sensory organs receiving stimuli from the environment and then converting this data into neural signals (Privitera, 2018).
3. **Attention:** Not every sensation leads to perception. Attention acts as a filter between sensation and perception, determining which stimuli we focus on. This can be influenced by a variety of factors including the intensity of the stimulus, our previous experiences, and our current needs or interests (Chun & Wolfe, 2008).
4. **Interpretation of Sensations:** Once the raw sensory data are captured and attended to, the raw signals are interpreted. Interpreting sensations involves attributing meaning to the sensory information, which is influenced by various factors such as our existing knowledge, past experiences, cultural background, and more (Goldstein & Cacciamani, 2021). It is through the application of attention and interpretation that the signal can progress to further processing and ultimately reach consciousness.
5. **Perception:** Perception is the process by which sensory information is organized and consciously processed, transforming raw sensory data into meaningful information. It enables us to comprehend and make sense of our surroundings. For example, perceiving and recognizing a dog involve the process of perception (Privitera, 2018; Mather, 2014). At this stage, a perception is not yet valued.
6. **Emotional Allocation of Perceptions:** Our emotions play a significant role in shaping our perception of experiences. The same event can be perceived differently based on our emotional state. An individual in a state of happiness

may perceive an event as joyful, while someone feeling sadness may perceive it as melancholic. This emotional allocation can impact our memory of the event and subsequent reactions (Dolan, 2002; Barrett & Bar, 2009).

7. **Experience:** Experience is a complex psychological phenomenon that encompasses the integration of sensory input, perception, and emotional reactions over time. Our experiences are subjective and unique, influenced by individual variables such as personality, mood, and cognitive processes like memory and attention. According to the definition provided by Qualinet (2013), an experience represents an individual's continuous stream of perception and interpretation of one or multiple events.
8. **Evaluation:** Evaluation is crucial as it involves comparing an experience against memorized or desired quality features. Through this comparison, we can derive a rating or assessment of the experience (Jekosch, 2005).
9. **Experience Rating:** In this model, experience rating refers to the final response or outcome, often represented on a rating scale.

The aforementioned steps are not discrete and often occur simultaneously or influence each other in a dynamic and interactive manner. Nonetheless, they provide a useful framework for understanding the complex process of forming an experience rating from exposure to a stimulus. However, there can be other elements to consider depending on the context and the individual involved: For example, cognitive appraisal (evaluating whether the perceived stimulus is a threat or opportunity), memory (extension of the stated "previous experiences" and "expectations"), social context (presence of others, their reactions, their influences), motivation and needs (e.g., if hungry, one's attention might be drawn to food-related stimuli, and they might rate food-related experiences more positively), or cultural factors (shaping values, beliefs, and expectations).

This model serves as the foundation for the development of the IAE model. However, the most important components for the understanding of the IAE model are highlighted in the model in Fig. 3.1:

- **Stimulus:** The stimulus itself plays a crucial role in estimating how it might be perceived.
- **Experience:** The experience is formed while consuming the stimulus and depends on various factors.
- **Individual:** The individual who has the experience brings their unique characteristics and traits into the equation.
- **Rating:** The final rating assigned by a person to the stimulus reflects their subjective evaluation of the experience.

While other components, such as sensation and perception, are relevant for understanding the process, they are not essential for developing the final model. Considering these four basic elements, there are two additional aspects to consider: the meaning of individuality and the impact of experiences beyond forming a rating.

In this work, individuality refers to all aspects that are unique to a person consuming a stimulus and providing a rating. This encompasses both time-invariant traits and momentary states (Mischel & Shoda, 1995). States can further be divided into those that depend on the stimulus-rating conditions, such as affective responses toward the stimulus being rated (Rauthmann et al., 2015), and time-varying states that are independent of the current situation, such as tiredness (Pilcher & Huffcutt, 1996).

The question of what happens with a live experience leads to the hypothesis that experiences influence behavior and physiological reactions, and that traces of experiences can be detected through the analysis of these reactions. There is evidence in the literature supporting the links between experiences and correlates with physiological and behavioral reactions, as discussed in Sect. 2.2.4. These findings suggest the existence of a "reaction" outlet of the modeled experience, through which additional aspects of the experiencing process can be assessed, even if the subjective evaluation (QoE ratings) does not capture these aspects. This is particularly interesting as physiological and behavioral dimensions allow for objective assessments that can be combined with a person's subjective responses (Brooks & Hestnes, 2010).

3.3 Structure of the IAE Model

The Individuality Assisted Estimation (IAE) model of experience, depicted in Fig. 3.2, elucidates how individual traits and temporal states influence a person's experience of a given stimulus, detailing how these influences can enhance the estimation of a user's assessment of the stimulus. As delineated earlier, the progression from *Stimulus* through *Experience* to *Experience Rating* is bolstered by three categories of individual traits, and by the *Reactions*, which encompass various physiological and behavioral measures.

3.3.1 Memory Through Feedback Loops

As delineated in Fig. 3.2, three entities in the model are linked with both inbound and outbound arrows. The outbound arrow signifies the effect of an entity on the *Experience* a person generates. This is a straightforward scenario where, for instance, a *Trait* such as openness to novel ideas or technology may impact the *Experience* an individual has in a new environment. Conversely, the model also accounts for reciprocal influences.

While it is proposed that an established experience dictates how someone evaluates that experience or responds to it physiologically, it is also hypothesized that these past experiences could shape future (e.g., emotional) states. Consider, for instance, a scenario where a person enjoys a high-quality call with clear video and speech signal, leading to a pleasurable conversation. This experience could engender a positive emotional state in the person. This positivity could potentially

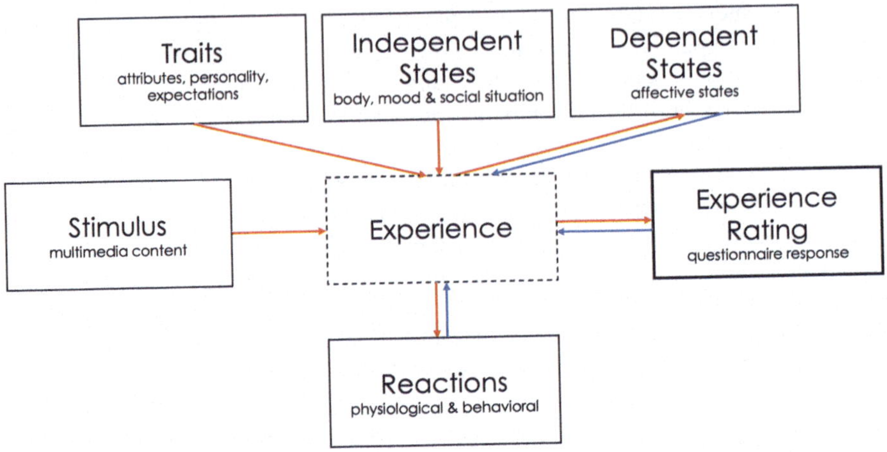

Fig. 3.2 The Individuality Assisted Estimation (IAE) model of experience. The model revolves around personal experience during the consumption of a multimedia stimulus, influenced by individual traits and temporary states. The latter are divided into situation-specific states and more general states, which vary over time but are independent of a particular stimulus. Together, these aspects impact physiological and behavioral reactions, shaping the perceived experience that ultimately determines a user's rating of the stimulus. To accurately predict a user's rating, the model integrates the stimulus itself, individual traits and states, and reactions (red arrows). It also considers feedback loops, hypothesizing that an experience-dependent state, for example, can influence subsequent stimulus experiences (blue arrows). This comprehensive approach enables the estimation model to leverage instrumental and highly individual observations, leading to a personalized estimation specific to each individual

color their perception of subsequent calls. While the quality of the ensuing call will undoubtedly be a significant factor in its appraisal, the residual positivity from the prior call could also impact the perception of the subsequent call. Similarly, the frustration arising from a poor-quality call could linger, subtly influencing the person's reaction to future calls.

To capture such carryover effects, the model includes three entities that can "remember" previous *Dependent States*, *Experience Ratings*, and *Reactions*. These entities possess a second arrow, extending from the core *Experience* entity outward, which serves as feedback loops. This structure enables the model to incorporate the influence of past experiences on present and future reactions.

3.3.2 Individual Traits

Traits encompass all aspects of an individual that are considered stable over time, unaltered by day-to-day changes, and not influenced by the stimulus or the overall surrounding environment. Examples include age group, mother tongue, and general neuroticism. Although these examples characterize certain aspects of a person, they are not dependent on the person's situation. Nonetheless, a certain population of a given age might be more accustomed to video-telephony than another, native speakers might comprehend heavily distorted speech signals more easily, and a highly neurotic person might struggle more with suboptimal call quality than someone with a comparably low score.

Consequently, the model acknowledges the potential influence of traits on an experience but does not presuppose a reciprocal relationship. Traits are observable through methods such as questionnaires or interviews and, since they are considered stable, this data acquisition is necessary only once for each participant.

3.3.3 Experience Independent States

Experience *Independent States* exhibit structural similarity to individual traits as they are specific to a person and independent of a current situation, stimuli, or quality experience. However, these states vary over time. Examples include tiredness or the intake of substances like cigarettes or drugs. A well-rested person might respond more energetically to a situation, while the same person lacking sleep may act sluggish and indifferent. Certain drugs may cause similar shifts in behavior or experience, and the consumption of cigarettes has been shown to affect physiological responses, such as Heart Rate Variability (HRV) (Dinas et al., 2013).

Similarly to traits, the model acknowledges the potential influence of experience independent states on experiences, but there is no backward influence on these states. These states are observable, for example, via questionnaires or interviews, and only need to be assessed once per participant.

3.3.4 Experience Dependent States

Experience *Dependent States* are flexible and dynamic. They depend on the current situation of a user and may change during an experimental session or be influenced by an experience. They are person-specific and highly variable over time. Examples of this category of states are affective states, mood and feelings, expectations toward a situation or task, thoughts and wishes, and even sensory adaptation (decrease in sensitivity of a receptor to a stimulus after constant stimulation (Privitera, 2018)). All these aspects are of a short-term nature and are greatly influenced by someone's experiences. Simultaneously, these aspects might influence how much attention is paid to a sensation, how this sensation is interpreted and perceived, how this perception is evaluated, and how it shapes a new experience. The model thus contains a feedback loop; an experience might influence the affective state, which in turn influences the subsequent experience.

Accordingly, the model illustrates a relationship from the dependent states to the experience, as well as from the experience back to the dependent states. Due to this sequence of influences, such states must be constantly monitored to be used in subsequent estimation processes. While the nature of the state dictates the investigative method, mostly questionnaires or short interviews would suffice to assess most states.

3.3.4.1 Reactions

The term "reactions" groups all responses that can be objectively observed. Physiological and behavioral responses can be assessed objectively using various sensors and observation techniques, with many even continuously observable through devices such as electrodes or camera recordings. As discussed in Sect. 2.2.4, there is substantial evidence suggesting correlations between cognitive processes, such as experiences, and physiological and behavioral features. For instance, highly stressful situations can objectively be measured through heart rate variability, as the organism transitions into a different physiological functioning when threatened. While it often seems more intuitive to posit that reactions follow from experiences, the relationship between reactions and experience is also modeled bidirectionally in the IAE model. For example, the concept of *constructed emotions* posits that physiological sensations and predictions from the brain engender emotions (Barrett, 2017). This challenges the traditional view that emotional states always follow cognition and instead proposes a reverse causality. Therefore, the model accommodates the influences of prior bodily sensations on experiences.

Most reactions should be measured as frequently as possible. Many of the reactions will be sampled at high frequencies and expressed through features derived from processing vectors of samples. Such operationalizations are discussed in Chap. 4.

3.3.4.2 Experience Ratings

The *Experience Rating* element represents the culmination of the model's estimations, encapsulating all relevant QoE metrics. It also incorporates a feedback loop to the *Experience* component to account for carryover effects from one rating to the next. It is hypothesized that the act of rating not only makes a person acutely aware of the situation but also initiates a reflective process. This heightened awareness could amplify their opinion to a degree that it becomes more memorable than it might have been without the act of rating. Consequently, this could result in an opinion carrying over to subsequent trials, influencing future ratings. Thus, this model component acknowledges and accounts for the potential impact of a previous rating on a subsequent one.

3.3.4.3 The Latent Experience Entity

The previously discussed elements of the model comprise an input stimulus that needs evaluation, three entities representing psychological individuality, a unit reflecting behavioral/physiological reactions, and the *Experience Rating* entity, which is the generated output by an individual for a specific audiovisual input. Additionally, the model features an intrinsic entity termed *Experience*. This latent and unobservable construct is intended to encapsulate all experience-related processes, which contribute to the formation of a rating when asked. An *Experience* thus signifies the personal involvement or interaction with stimuli or situations that influence and mold one's perception, cognition, emotions, and viewpoints. Consequently, a rating is an outcome of an assessment process related to a specific dimension (e.g., speech quality), drawn from the *Experience* one is undergoing at the time of forming a rating. This distinction is depicted in Fig. 3.1 with its additional *Evaluation* entity. A rating is not formulated until a particular *Experience* is assessed and a rating is developed. The IAE model includes the inner *Experience* entity to differentiate between the *Experience* itself and a rating derived from it, bridging all other entities.

3.4 Validation Methods

The proposed model sets forth several hypotheses in the form of involved elements and the edges connecting them. The directional nature of these edges implies that the hypotheses are directed as well. To validate the structure of the model, three types of analyses are carried out in the subsequent chapters:

- Appropriate data sets are explored to find evidence for each of the edges. This step aims to validate each individual hypothesis. The strategy involves conducting numerous correlation and group tests to ascertain if *any* of the variables of one model entity show a statistically significant relationship with *any* of the variables or features of the connected model entity. Because the model does not specify particular types of analyses, sensors, or dimensions to represent the model entities, multiple variables and instruments are involved. Therefore, in the experiments introduced in the following chapters, various aspects of each entity are examined. For instance, *Traits* is not a construct that can be measured with a single scalar value. Instead, psychology suggests, for example, five personality dimensions. Each of these dimensions are potential candidates for the proposed correlation analysis. If even one of these correlations proves to be statistically significant, valid, and reliable, this provides evidence for the model's edge that suggested such correlations. Apart from testing all edges, this approach also yields a list of relevant variables. As part of the validation process, many potential variables are considered, and not all of them are expected to achieve statistical significance. Consequently, narrowing down the list of options is also a desirable outcome of this step. This type of analysis is carried out and discussed in Chap. 5.

- The structure of the model suggests a Structural Equation Model (SEM), as this statistical technique accounts for different influences and allows for computing multiple regressions simultaneously. This method enables the entire structure to be modeled at once, providing estimates and significance values for each variable involved. While the initial correlation approach might identify several variables as having a statistical relationship with each other, this may change when all influences are modeled simultaneously. This is due to the possibility of multiple variables correlating with each other and explaining the same share of variance. To identify such instances and to test the edges of the model with a unified approach, an SEM-analysis is presented in Chap. 6. After these two analyses are conducted, the model's structure is considered to be empirically validated, provided the data does not suggest any structural changes.
- Lastly, the model is applied to perform estimations. Ultimately, the model aims to improve estimations of perceived QoE and to predict individual ratings more accurately than traditional approaches, such as signal-based ones. This practical application and the analysis of predicted results represent the final type of analysis conducted to validate the model. If the IAE model of experience yields better results than its signal-based predecessors, it demonstrates its utility and effectiveness, thereby validating the outcomes of this research. This is presented in Chap. 7.

3.5 Estimations and Applications

The successful validation of this model would provide a robust and nuanced framework for predicting perceived QoE. Therefore, all individuality related elements of the model would serve as additional information for an estimation system. Traits and states, as well as reactions, would feed additional user information into an instrumental quality estimation model, that not only operates on this personal data, but augments existing, objective estimation using these individuality information. Figure 3.3 visualizes this individuality informed estimation approach.

Evidence supporting the assumption of an increased estimation accuracy by adding knowledge about a user into the estimation system would be beneficial in several ways:

- **Holistic Understanding of QoE:** The model takes into account various factors that influence QoE, providing a more comprehensive understanding of the dynamics at play. It not only considers signal quality but also other elements like individuality, attention, and emotion.
- **Improved Predictive Accuracy:** Traditional models often rely on signal-based factors for predicting QoE. However, these approaches might fail to account for important factors such as viewer's individual characteristics and states, which

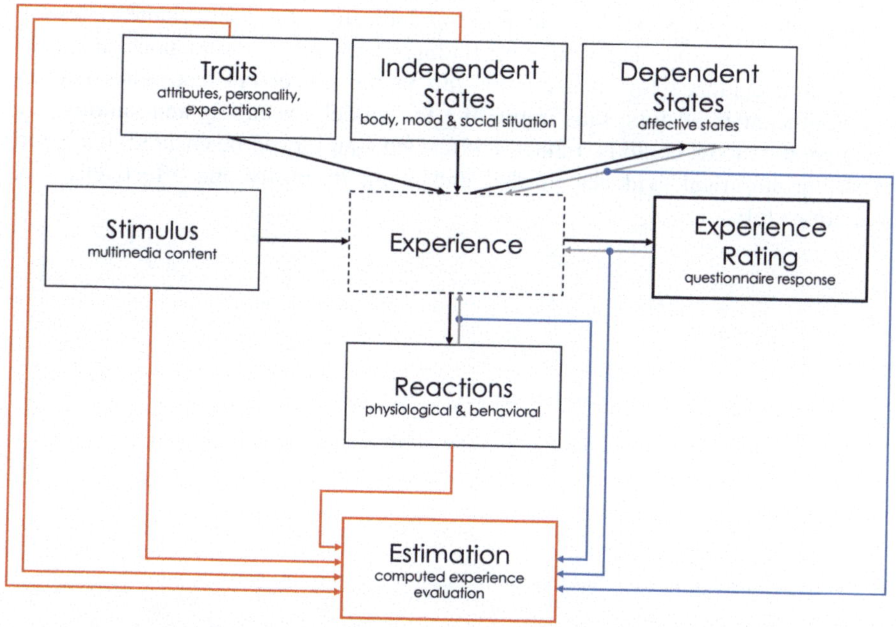

Fig. 3.3 The configuration structure of the complete level of the Individuality Assisted Estimation (IAE) model of experience, illustrating the process of individuality informed QoE estimation. In this visualization, pathways are added from the signal source and all individuality aspects to an estimation unit. The estimation unit incorporates an objective stimulus score as well as reactions and the three psychological individuality aspects distinguished by the IAE model (represented by orange pathways). Feedback loops are also tapped (through the blue arrows), capturing subjective or objective measures about the user from a preceding trial and feeding them back into subsequent trial experiences. This enables the model to recognize changes in the user that may be influenced by a stimulus and subsequently impact future experiences. This structural configuration gives rise to the name of the model: The Individuality Assisted Estimation (IAE) model of experience, which enhances the estimation process by incorporating additional information about each specific individual

the proposed model considers. Thus, it is expected to improve the accuracy of estimations.

- **Guidance for QoE Enhancement:** This model provides valuable insights into the impact of factors such as attention and emotional allocation of perceptions on QoE. By leveraging this understanding, it can offer guidance for strategies aimed at enhancing the quality of multimedia services. This guidance holds potential benefits for network providers and Over-the-Top (OTT) services, as it would enable them to dynamically adjust network resources based on, for example, subjective demand, thereby improving the overall User Experience (UX).
- **Versatility and Generality:** The model is agnostic to specific types of analyses, sensors, or dimensions. This means it can be used flexibly across different contexts and applications, making it highly versatile.

The outlined steps, such as identifying indicators for each element, testing responses in isolation, and validating findings with real conversations, not only aim to find evidence for each relationship in the model but also aim to reject non-significant indicators, thus improving the model's accuracy and validity. By doing so, the model could be refined—if needed—and it can be ensured, it's based on strong empirical evidence, further enhancing its utility and effectiveness in predicting QoE.

Chapter 4
Operationalization of the IAE Model Entities: Relevant Dimensions in the Context of this Work

Experience without theory is blind, but theory without experience is mere intellectual play.
Immanuel Kant

The following chapter adheres to Immanuel Kant's assertion; thus, this chapter lays out the architecture of the proposed IAE model, anchoring it firmly in empirical data. As previously examined in Chap. 2, the QoE domain presents an extensive array of metrics and measures. However, a concise and purposeful subset is required to evaluate the research questions delineated in Sect. 2.4.

Quality perception, especially concerning video-telephony speech quality, video quality, and audiovisual quality, constitutes the core metrics for quantifying QoE. But there are other experience-related concepts warranting exploration as well. The objective is to construct a comprehensive depiction of human experience within a multimedia context, using a compact set of dimensions. The following dimensions are selected:

1. Quality perception: **audio**, **video**, and **audiovisual quality**, crucial in the video-telephony context.
2. Cognitive dimensions: **mental load** and **effort**, describing the degree of cognitive exertion during a task.

The mental load dimension is intriguing as it gauges how mentally taxing a situation is for an individual. Existing literature supports the notion that poor quality necessitates higher mental exertion, while superior quality can reduce cognitive demand for the same task. The NASA TLX questionnaire is employed after each trial to assess mental load and its subdimensions. Effort, similar to mental load, measures the exertion needed to consume a multimedia artifact. In addition to the "effort" item included in the NASA TLX, the ITU's listening effort scale is utilized.

Emotional experiences, an integral aspect of human perception of a situation or stimulus, are also considered. However, as detailed in the following chapter, emotions are evaluated as a transitory state than an outcome measure. This is because the primary goal is not to develop a method for estimating emotional

R. P. Spang, *Individualized Quality of Experience Estimation in Audiovisual Communication*, T-Labs Series in Telecommunication Services,
https://doi.org/10.1007/978-3-031-67551-5_4

responses but to focus on perceived QoE. As Sect. 2.1 defines, "QoE is the degree of delight or annoyance of the user . . ." (Qualinet, 2013). As such, emotional states are a crucial influencer. This is captured by the model through its *Dependent States* entity and especially through its feedback loop.

The chosen subset for describing experience should encapsulate the most relevant aspects of multimedia consumption. Apart from quality perception, it also includes cognitive dimensions such as effort and load. This provides a foundation for evaluating subjective experiences, and the remainder of this work will concentrate on these dimensions, although further dimensions of human experience might warrant exploration.

The IAE model of experience suggests four observable entities besides a stimulus and the QoE dimensions: Individuality, Dependent and Independent Temporal States, and Responses. These entities focus on the participant, their background, and their physiological and behavioral reactions to a stimulus. There could be numerous variables and dimensions to consider for each entity. As discussed in the Fundamentals chapter, existing literature provides evidence for many potential variables. However, to the best of my knowledge, there is no systematic approach to select the most meaningful variables for quantifying human experience.

The selected subset includes various categories of variables that collectively portray a comprehensive picture of a person's identity, their current emotional state, and their reaction to a given situation. Yet, there's no systematic way to confirm that no significant variable is missing. The proposed solution to this potential issue is to systematically assess if the selected variables show a meaningful relationship with the other dimensions or entities of the IAE model. This evaluation is the subject of the following Chap. 5.

4.1 Stimulus Manipulation and Experimental Paradigms

The experimental data for this work are based on two different experimental paradigms: a passive conversation simulation and an active conversation study. These two experiments have both been replicated several times, but both cases followed the same procedure, structure, stimuli, or manipulations, so the data of several studies can be pooled. The studies themselves are further described in the respective sections of Sects. 5.1 and 5.2.

In order to create realistic degradation, a target space for quality degradation has been defined. Other than multidimensional quality spaces, e.g., ITU-T Recommendation P.918 (2020), the aim was to label video clips in simple categories. Through qualitative expert interviews, four plus one essential classes of video-telephony quality have been identified (Spang et al., 2022b):

1. No impairment, connection maintained in best quality
2. Throttled bit rate, reduced quality for more stability
3. Discontinuity I: continuous, short dropouts and delays

4. Discontinuity II: connection is mostly frozen
5. Connection terminated prematurely

While this is a broad definition, it provides a common ground for discussing different severities. It must be noted that while this list captures all stages from perfect to canceled, only the first four stages can meaningfully be simulated. The fifth stage, cancellation, could be added using a black screen. Both studies aligned to these essential quality categories; both presented stimuli (or the conversation) in four different degradation conditions, resembling the categories one to four.

4.1.1 Video-Based Conversation Simulation

To study physiological and behavioral responses to quality changes, an experimental environment was needed to record subtle signals on the participants' side and to control various influencing factors on the experiment's side. The proposed solution was to utilize conversation simulations, restricting potential confounders and controlling variables such as social aspects or speaker voices by using prerecorded conversation elements. Real-world conversations introduce noise due to human social factors, making the simulation approach essential for consistent and controlled stimuli presentation. The primary goal was to ensure that the presented stimuli were replicable, reflecting genuine conversations and demonstrating quality variations. Following these criteria, a benchmark dataset was created for groundwork and further research purposes. The dataset is publicly available under a CC-By 4.0 license.[1]

This approach was first detailed by Belmudez et al. (2013) but has not been previously used to isolate behavioral and physiological signal changes due to quality degradation. In previous studies, pre-recorded samples were used to simulate conversations for quality research purposes, but their stimuli material was often not published. ETSI Technical Report 102 506 (2007) and Weiss et al. (2009) have described a conversation simulation test, where they compared real MOS ratings to model-based quality rating estimations using pre-recorded telephone samples. Others used voice or movie scene–centered datasets, which were not tailored for everyday calls, further highlighting the need for more specific, conversation-focused stimuli.

To fill this gap, the Storytime dataset was created, composed of 160 video clips divided into four narratives, each with ten interrelated parts rendered in four degradation levels. These narratives maintain a common thread with the use of the same actor, ensuring content diversity while maintaining uniformity in presentation. Each narrative was designed to engage viewers by incorporating questions related to the actor's spoken content and visual elements. These clips were recorded using

[1] https://osf.io/cyb8w/

Fig. 4.1 Example scenes of the different stories, their settings, and costumes. From top left to bottom right: "Dinner", "Holiday", "Hiking", and "Office"

popular video conferencing software (Zoom and Skype), providing high ecological validity by replicating actual artifacts found in real-world video conferencing. This dataset is especially useful for experiments requiring consistent conditions, such as those exploring physiological responses. All dataset details and an evaluation user study are included in Spang et al. (2022b).

The design of the experimental setup and the use of recorded material were significantly informed by the studies conducted by Weiss et al. (2009) and Belmudez et al. (2013). The ITU-T Recommendation P.911 (1998) guidelines formed the basis for the duration of the videos, with each video designed to last between 8 and 12 seconds.

Figure 4.1 showcases a collage of scenes from each narrative. The videos were designed to mimic typical video-telephony environments and were recorded using a Sony A7C camera and a Rode SmartLav+ microphone. They are available in FullHD resolution (1920×1080), 29.97 fps, and encoded in H.264 format within an MP4 container, with audio encoded as AAC at 48 kHz and 32bit resolution.

To generate realistic distortions, a network router around the Linux kernel module NetEm was built to channel network traffic through a manipulated network interface. This allowed me to manipulate parameters such as bandwidth, delay, jitter, and packet corruption. Video calls were recorded using Zoom v.5.9.3 and Skype v.8.80.1 to produce genuine distortions between two clients in the manipulated network. By streaming the pre-recorded clips through these platforms using Open Broadcaster Software (OBS) and screen-recording the clients on the receiving end, I created realistic distortions of the video database.

Initially, the goal was to identify specific network conditions that could distort videos corresponding to a given degradation level. However, I found that these video-telephony clients were resilient under severe network constraints, prioritizing speech signals over video quality. To overcome this, I frequently changed network settings during the recording process. As a result, each clip was recorded multiple times, producing different quality versions.

In the final stage, the screen recordings were evaluated using instrumental quality estimation models. This algorithmic approach was used to select the best matching recording for each clip and each of the three quality categories (categories two to four as per listing 4.1). This approach ensured high ecological validity, closely resembling real-life distortions encountered in productivity systems like Zoom and Skype. Thus, every original clip was distorted multiple times to match a predefined range of MOS values and other dimensions per degradation levels. This methodology resulted in a diverse and realistic dataset that represents an array of distortions encountered during video communications. See Spang et al. (2022b) for further details.

4.1.1.1 Task

The primary task was divided into three blocks, each consisting of ten trials. Each trial featured a logically sequenced part of a story; however, the order of the four different stories was randomized. Following each video, a question regarding visual or acoustic information from the previous clip was presented. Under optimal quality settings, each question can be answered. However, some videos of degradation levels 3 and 4 were significantly distorted, making these few content-questions unanswerable.

The questions served a dual purpose. firstly, to encourage participants to pay close attention and engage fully with the video's content, thereby enhancing the realism of the simulated conversation. Secondly, the responses to these questions provided an indication of content comprehension. In this way, correct answers functioned as a quality metric of the clips, encoding intelligibility independent of the MOS ratings. This mimics a real conversation where participants might struggle to follow along under heavy degradation conditions.

Subsequently, participants were asked to rate the overall audiovisual quality of the clip, as well as the video and speech quality separately. After each block, participants were asked three questions relating to story likability, speaker engagement, and speaker likability, rated on continuous scales.

4.1.1.2 Procedure

Before each session, the simulated nature of the experiment was emphasized to the participants. They were asked to visualize a video-telephony situation with a friend, recounting their activities over the past few weeks. The experiment itself was conducted in a lab room at Technische Universität Berlin, under controlled lighting and environmental noise conditions, following ITU-T Recommendation P.911 (1998). The distance between the participant's head and screen was kept consistent

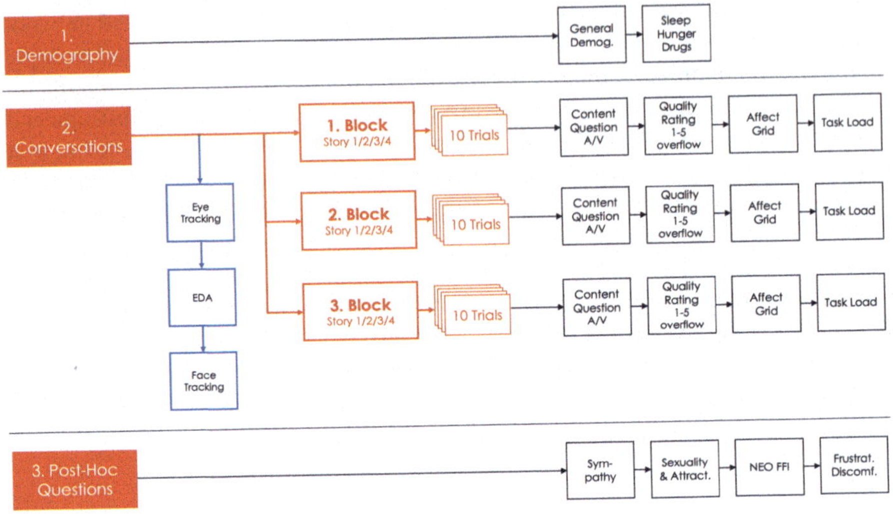

Fig. 4.2 Procedure of the video-based conversation simulation experiment. In total, three video stories, with ten clips each, were shown to the participants. After each trial, questions regarding the perceived quality, valence and arousal, and perceived task load were asked. Physiological and behavioral measures were recorded over the entire time of the experiment

for all participants. Participants were seated at a standard office desk equipped with a 24-inch screen, a standard keyboard, and an office mouse. Participants used headphones (over-ear, AKG K601) to listen to the audio signals of the videos. The study procedure, programmed in PsychoPy Version 2022.1.1, displayed the videos and subsequent questionnaires after each trial, storing timestamps for each event to synchronize with sensory data recorded simultaneously.

The order of the stories, as well as the degradation levels of the clips, was randomized. Two consecutive clips shared the same degradation level, meaning the first and second clips were of the same degradation level, followed by the third and fourth of a different degradation level, and so on. After completing the main task, participants received compensation for their time. The procedure of the video-based conversation simulation experiment is displayed in Fig. 4.2.

4.1.2 Conversation Paradigm

ITU-T Recommendation P.805 (2007) forms the basis of the conversation experiment setup (see Figure 6–1 of ITU-T Rec. 805). In the present case, a WebRTC connection was established using a communication platform developed for these experiments. Here, a client A was connected to a call-server C. From this call-server, a second WebRTC connection was established to the second client B. Participants sat in two different rooms in front of clients A and B. The connection quality was

Table 4.1 Network manipulations for the conversation call-server to generate four distinct degradation conditions ("degrad. lvl") of the WebRTC connection

Degrad. lvl	Target quality description	Delay	Packet-loss
1	No impairment, connection maintained in best quality	0 ms	0%
2	Throttled bit rate, reduced quality for more stability	500 ms	7.5%
3	Discontinuity I: continuous, short dropouts and delays	0 ms	40.0%
4	Discontinuity II: connection is mostly frozen	1000 ms	60.0%

altered by manipulating the network adapter on the call server. Here, all outgoing packets were affected by the network manipulation, which was realized using *dnctl* to create a dummy network with the desired parameters, and *pfctl* to route all traffic through the dummynet. Both are part of a standard Apple macOS installation; here macOS Monterey version 12.6.5 was used for the experiments. The double-connection design was chosen to force the WebRTC-stream to use the call server as a hop in the middle and not to directly connect the two clients peer to peer. Via the call server's manipulated network, the connection quality of the WebRTC-connection was reliably altered, again with the four degradation levels defined in listing 4.1 in mind. Table 4.1 details the connection manipulation used to create four degradation conditions of the experiments. The high packet-loss rates were necessary to create a meaningful distortion, since modern-day packet-loss concealment techniques work very well. The objective here was focused on creating clearly distinct degradation conditions.

4.1.2.1 Task

The experimental procedure begins by instructing the two participants to engage in two distinct tasks—each correlated with different levels of network performance. These tasks include a survival task and a story comparison task.

In the survival task, participants are presented with a hypothetical survival scenario, such as an accident in a remote location. Each participant receives a unique list of items, with some overlap between lists, and they must discuss and reach a consensus on which four items are most essential for survival within the given scenario. This task is designed to stimulate natural conversation and decision-making processes. More details on the item lists and scenarios can be found in ITU-T Recommendation P.1301 (2017).

In the story comparison task, each participant is given a short story to read beforehand. These stories are similar but contain subtle differences that participants must identify through conversation. Participants are provided with guiding questions to help them recall critical aspects of their stories, ensuring that their focus remains on the audiovisual channel during the discussion. The goal is to discover all differences between the stories as quickly as possible without consulting the text during the conversation. Detailed descriptions of the stories and guiding questions can be found in the original ITU-T Recommendation P.920 (2000).

Refer to Appendix V.4 of ITU-T Recommendation P.1301 (2017) for the picture material used in the survival task trials. The images originate from the ITU-T recommendation, and their description text has been translated into German. Appendix A.1 of this work contains the German stories provided to the participants for the story comparison task.

While the survival task encourages relatively unrestricted conversation, the story comparison task prompts a more focused discourse.

4.1.2.2 Procedure

The experiment was conducted in overall three lab rooms at Technische Universität Berlin, under controlled lighting and environmental noise conditions. Both participants sat in a room on their own, while the experimenter was overseeing the progress of the experiment at their control position next to the call server. Participants were seated at a standard office desk equipped with a 24-inch screen, a standard keyboard, and an office mouse. Participants used headphones (over-ear, AKG K601) to listen to the audio channel of the conversations. The study program was a webapp, developed in vue.js and powered by a node.js backend, the calls were realized as WebRTC connections in full screen mode. Before each call, the upcoming call situation was briefly explained on screen. After each call, an on-screen questionnaire appeared, which the participants answered using their mouse and keyboard. Each step of the experiment had to be invoked via the click on a "Ready" button. This button was synchronized among both clients, so that the experiment only continued if both participants clicked ready. This ensured that both parties were always at the same stage of the experiment.

The order of the degradation conditions was randomized, although the order of the two tasks was always the same. After completing the main task, participants received compensation for their time. The procedure of the video-based conversation simulation experiment is displayed in Fig. 4.3.

4.2 Experience Ratings

The following dimensions are considered relevant experience dimensions in the context of this work. The quality-related questions are probably the most straightforward QoE dimensions. However, also the necessary effort one has to put into understanding each other, as suggested by the ITU, is considered. In the same vein, also the cognitive task load a person felt was required during a call situation is included into this selection. These dimensions were assessed after each trial of all experiments presented in the following chapter, so ratings are obtained for every call once.

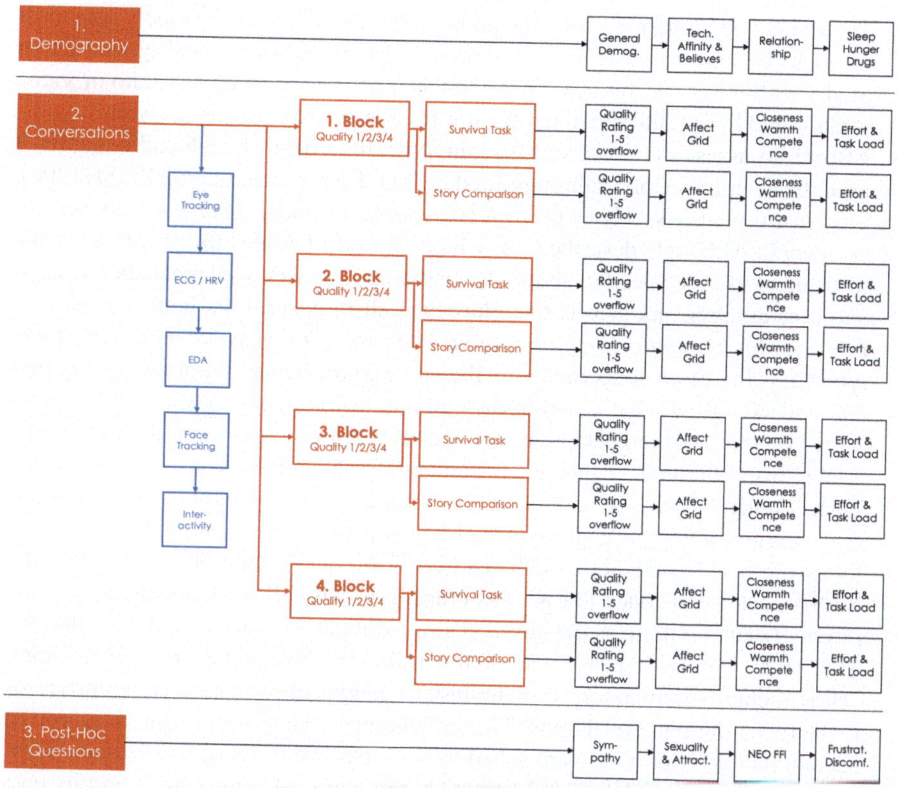

Fig. 4.3 Procedure of the real conversation simulation experiment. In total, four blocks á 2 trials were recorded. Each block was set to one of four degradation levels. Within each block, participants engaged in two tasks: a survival task and a story comparison task. After each trial, questions regarding the perceived quality, valence and arousal, perceived task load, and the social relationship between the two participants were asked. Physiological and behavioral measures were recorded over the entire time of the experiment

- **Quality ratings (audiovisual, speech, visual):** To assess the perceived quality of the video-telephony connection, an extended continuous scale (seven-points, including overflow), as per ITU-T Recommendation P.804 (2017), with German labels, was used (Bodden & Jekosch, 1996). Quality ratings with this scale have a strong relation to the standard ACR scale and improved sensitivity, especially on the lower end (Köster et al., 2015). The scale is added as Appendix B.1. After each trial of all experiments, these three aspects were always the first dimensions to ask the participants to rate. The overflow scale was used as pre-tests of experiments described in the following chapter revealed flooring effects in the worst degradation condition. To overcome these, the overflow scale helped normalizing the obtained ratings. Appendix B.1 contains the scale used in all experiments. The focus of this work is to improve the estimation of the audiovisual quality using information about the individual during a call situation.

The rationale behind recording the audio and video channel separately was most of all to be able to correlate the scores of an instrumental quality estimation model with the users' ratings. Such models mostly process either audio or video steams, and hence, the separate user impressions were collected as well.

- **Effort:** To assess the perceived listening/viewing effort of each call situation, I used the listening effort scale suggested in ITU-T Recommendation P.800 (1996). This instrument asks about the *Effort required to understand the meanings of sentences*. Although described as a listening-effort scale, the wording of the instrument is not specific about the audio or the video channel, but tests if and how cumbersome it was to follow the call in general. Its items range from *full relaxation, no effort required*, to *no understanding possible with reasonable effort*. It is therefore concerned with the *effort* someone has to put into taking part and understanding what's said in the call and to follow the task at hand. Higher scores on the scale indicate a greater necessity for effort, or even the participant feeling overwhelmed, suggesting a challenging call situation.
- **Task Load:** While the task load is related to the previous *effort* scale, it is concerned with the cognitive load a given task (here: participating and or following a video-telephony call) caused in the participant. To assess this, the NASA Task Load Index (TLX) was employed, with all its items apart from the *physical demand* item, as the physical strain did not vary between the conditions. The TLX score is the average of the following five items: *mental demand*, "How mentally demanding was the task?"; *temporal demand*, "How hurried or rushed was the pace of the task?"; *performance*, "How successful were you in accomplishing what you were asked to do?" (reversed item); *frustration*, "How insecure, discouraged, irritated, stressed, and annoyed were you?"; and its own *effort* item, with the descriptive questions "How hard did you have to work to accomplish your level of performance?" (Hart & Staveland, 1988). The questions were presented in German (Flägel et al., 2019). The TLX score ranges from 0 to 20, whereby a score on the upper end means that a person had perceived a high or intense task load.

While there are other dimensions that could be interesting to consider as relevant for a user's experience, for example: emotional states, this work focuses on the stated three dimensions. In the case of emotional states, these are considered as stimulus-dependent, individual states.

4.3 Dependent States

Dependent states are, according to the model introduced in Chap. 3, those psychological or mental states that are influenced by a recent situation or experience. They not necessarily have to depend on the quality that was observed just before and may have different sources. However, they are united in that they are influenced and can be altered within a relatively short period of time. The IAE model suggests that previously made experiences influence these states, and in addition, these states influence the current experience back. This way they serve as a memory of what happened, and more importantly, how a person felt before the current experience.

The evolutionary sciences introduced the term "carry-over effects" for expressions of trait values in an earlier life stage that are directly carried over into subsequent life stages (Moore & Martin, 2019). The IAE model suggests similar effects for the experience formation process. Practically, these always refer to the preceding trial. The literature offers plenty of studies demonstrating influenced experiences through previous states. For example, priming studies fall in this category: Harrison et al. (2013) found evidence that affective priming significantly influences visual judgments, and that positive priming increases performance. Another example is mood induction, in which certain mental states are artificially evoked (e.g.: Baños et al., 2012). Together with the literature of how different moods influence quality perception (e.g., Reiter et al., 2014), the case is to be made that the state, how someone feels, before exposed with a multimedia artifact of a certain quality, influences the person's quality perception and their experience.

Time is an important aspect when discussing such carryover effects, as they fade away rather quickly. Arguably, there are many layers to that, ranging from very short time to longer lasting ones. However, this is not in scope of this book and only one previous experience made is considered for the dependent state analyses of the following. This is done by assessing the following ratings after each trial and then testing if there are any influences of the previous ratings of these states, onto the experience ratings of the subsequent trial. This is described in more detail in Sect. 5.1.2.3.

4.3.1 *Affective States*

To capture the emotional response toward a recent call situation, Russel's affect grid was employed to assess the dimensions **valence** and **arousal** (Russell et al., 1989). Both together form an affective response and capture how positive a situation was, ranging between *unpleasant* (low value) and *pleasant* (high value), and to capture how arousing or exciting the situation was, ranging between *Sedation/Sleepiness* (low value) and *strong arousal* (high value). These values were assessed using a two-dimensional grid in which participants clicked once to respond to both dimensions. The x-coordinate encoded the valence part, while the y-coordinate encoded the arousal part. This transformation returns continuous values for both dimensions.

The idea behind the choice to add this to the dependent states is, that if a person feels very well, it is unlikely that a bad audiovisual quality will fundamentally change the positive affect. It is reasonable to assume that the stimulus' quality alters a person's affect to a degree. But it is likely that the state one enters a situation with influences the evaluation of the situation. Put differently, this is the hypothesis that human ratings are not state-independent, but are influenced by the state someone is in before the evaluation process starts.

4.3.2 Preference to Abandon the Call

To evaluate how individuals might respond to a specific quality degradation, participants were asked after each trial whether they would prefer to disconnect and reattempt the call or maintain the connection if there were more topics to discuss. Their responses were measured on a continuous scale, with 0.0 indicating that they would choose to end the call *(I would not continue the call and hang up)* and 1.0 suggesting that they would continue the conversation *(I would continue the call)*. This measure provides insight into the *preference to abandon* the call, simulating how they might react if such quality degradation occurred in a real-world scenario.

It is important to clarify that during the conversational experiments, participants could end the call at any moment through a provided *hang-up* button on the interface. Even though the call was automatically terminated after 5 minutes, participants were free to hang up earlier if they believed they had resolved the problem, if they found the quality unbearable, or for any other reason. However, given the encouragement to complete the task and the somewhat artificial lab environment, it is anticipated that most participants would persevere. Therefore, their hang-up behavior in this setting might not accurately reflect their usual, personal preferences.

Consequently, the *preference to abandon* measure can offer valuable insights into their attitudes under different degradation scenarios.

4.3.3 Social Perception

To measure the **closeness** of a bilateral social relationship between individuals, Aron et al. (1992) developed the Inclusion of the Other in the Self (IOS) scale. The IOS scale is a simple, pictorial tool that is easy to use and highly intuitive, making it highly appropriate for repetitive measurements of introspective judgments on a low conscious level. The scale is neutral about the reasons why people feel close or distant. It measures their sense of being interconnected which each other, which may be the result of conscious or unconscious processes.

The perception of a human vis-à-vis is an evolutionary reflex to quickly assess the surrounding environment for danger. Positive or negative moods affect attitudes and behavior, as shown in research on the topic of affective priming (Fedorikhin & Cole, 2004). To simplify this process of social perception, it is dominated by social cognition using social stereotypes (Fiske, 1993). In this sense, social cognition is the unconscious attitude toward another person or a social group (Fiske, 1993). Fiske et al. (2002) found two basic dimensions of social cognition: **warmth** and **competence**. These dimensions follow the two automatically running questions, and these are which aims does the other person pursue and how effective will he or she be in achieving these aims (Cuddy et al., 2008). In the context of this work, the two dimensions are used to capture the social perception of an interlocutor.

Previous research in the multimedia quality domain has established the relationship between quality and the perception of an interlocutor or presenter (Newman & Schwarz, 2018), so an influence of the call quality on the interlocutor's warmth and competence is hypothesized and the reason to add these dimensions to the list of dependent states to be investigated. For both dimensions, a continuous scale was used. The question was *To what extent do the following descriptions apply to your interlocutor during the last conversation?"*, then both scales were titled with either *warmth* and *competence*, and both ends of the scales were labeled with either *Does not apply* or *Does apply*.

Preliminary studies have indicated that these dimensions can exhibit variations under different degradation conditions. This variability is particularly pronounced among participants who are not acquainted prior to the conversation experiment. Consequently, the construct has been incorporated as one facet within the *Dependent States* entity, recognizing its role in shaping subjective experiences.

4.4 Independent States

Independent states are, according to the IAE model, such psychological or mental states that do vary over time, but are independent of a given stimulus or a multimedia artifact. This is what distinguishes them from the dependent state. Hence, the model does not assume any changes of the independent state, based on various experiences someone has. Because of that, the following dimensions are collected only once, before the main experiment.

4.4.1 *Social Relationship and Sympathy*

Regarding the social relationship, three questions were asked:

- **Relationship intensity**: assesses how strong a person evaluates their relationship with their interlocutor, compared to all their other relationships: *Wie würdest du deine Beziehung zu deinem Gesprächspartner im Vergleich zu all deinen anderen Beziehungen (sowohl gleich- als auch andersgeschlechtlich) charakterisieren?* The rating is recorded through a discrete, seven-point rating scale. The left end is labeled *Überhaupt nicht nah* (not at all close), while the right side is labeled *Sehr nah* (very close). Unlike social perception, this item evaluates the overall relationship, irrespective of any specific situation.
- **Familiarity**: assesses the category of relationship the person has with their interlocutor. The question is simply *Do you know your interlocutor?*. Options to choose from are *Nein, ich kenne sie/ihn nicht* (no, I don't know her or him), and *Ja, wir sind:* (yes, we are:), followed by a list of options, whereby multiple elements could be selected: *Lebens- oder Ehepartner* (life partner or spouse),

Familie (family), *Freunde* (friends), *Bekannte* (acquaintances), and *Kollegen* (colleagues).
- **Have had calls**: assesses if the two have had video-telephony calls with each other in the past: *Hast du mit deinem heutigen Gesprächspartner in der Vergangenheit schon einmal videotelefoniert?* If they never met before, this is obsolete, but it is interesting to assess if they are used to calling each other using video-telephony and have made experiences of such situations before, because, for example, many family members never had a video-telephony call with each other before. This item was recorded binary.

Sympathy for each other was assessed once after the main part of the experiments, after participants talked to each other. The sympathy complex consists of four items, each was assessed via a continuous scale, labeled *stimme überhaupt nicht zu* (completely disagree) on the left-hand side and *stimme voll und ganz zu* (completely agree) on the right-hand side.

- **Would meet again**: The first item asked if the participant would like to meet their interlocutor again: *Ich würde mich gerne wieder mit meinem:r Gesprächspartner:in treffen.*
- **Could be a friend**: The second item asked if the interlocutor could be a friend of the participant responding to the question: *Ich denke, mein:e Gesprächspartner:in könnte ein:e Freund:in von mir sein.*
- **Is pleasant**: The third item asked if the interlocutor is a pleasant person to talk to: *Ich denke, mein:e Gesprächspartner:in ist ein:e angenehme:r Zeitgenosse:in.*
- **Circle of friends**: The last item asked if the interlocutor would fit into the circle of friends of the participant: *Ich denke, mein:e Gesprächspartner:in würde in meinen Freundeskreis passen.*

4.4.2 Hunger, Sleep, Drugs, and Cigarettes

Furthermore, biological or chemical influences were also assessed. Here, I focused on the following four elements to create a reasonable variety of different, possible influences:

- **Hunger**: In a recent publication, colleagues and I demonstrated hangriness (hunger and anger) in audiovisual quality perception (Spang et al., 2023c). To test for hunger based influences, two questions were asked; if the participant feel hungry right now, *Hast du gerade Hunger?*, assessed on a continuous scale from *Trifft nicht zu* (does not apply) to *Trifft zu* (does apply), as well as how long ago did the participant eat something for the last time, *Vor wie vielen Stunden hast du zuletzt etwas gegessen?*. For the latter, an input field was offered in which participants entered the past time in hours since they last ate.
- **Sleep**: Similarly, questions regarding sleepiness were assessed: First, if the participant feels tired right now, *Bist du gerade müde?*, similarly on a continuous

scale, how many hours did the participant sleep today, *Wie viele Stunden hast du heute geschlafen?*, for which a similar input was used to enter the number in hours, and additionally, how many hours the participant usually needs to feel well rested, *Wie viele Stunden Schlaf benötigst du üblicherweise, um dich fit, wach und ausgeschlafen zu fühlen?*, assessed through a similar input field. Hunger and sleep might both directly influence the affective states people perceive.

- **Drugs**: The next two items were assessed to explain possible influences on physiological responses. Regarding drugs, we asked if any mood altering medication or substances were consumed today, *Hast du heute stimmungsverändernde Medikamente oder Substanzen eingenommen (z.B. Antidepressiva, Neuroleptika, Cannabis, etc.)?*, with a binary response option: *Ja* (yes) and *Nein* (no).
- **Cigarettes**: Regarding cigarettes, we asked if the participant does consider themselves as a smoker, *Bist du Raucher:in?*, with a binary response option: *Ja* (yes) and *Nein* (no). If they indicated so, a second item was offered to indicate the number of cigarettes smoked today, *Wenn ja: Wie viele Zigaretten hast du heute geraucht?*. This number was assessed through an input field. While being a smoker is rather a demographic information than a state, it is grouped together here. Smoking has shown to influence several physiological dimensions; hence, it is relevant in this context.

4.5 Traits

4.5.1 OCEAN Personality Dimensions

The "Big Five" personality traits: neuroticism, extraversion, openness to experience, agreeableness, and conscientiousness (Goldberg, 1990). These five factors that should systemize the enormous number of characteristics in which people reliably differ (Schnabel et al., 2002). For example, individuals with high neuroticism may be more sensitive to low-quality multimedia content (Scott et al., 2016). In the context of this work, the *NEO-FFI* questionnaire was employed, once per participant (Costa & McCrae, 1989).

4.5.2 Frustration Intolerance

The Frustration Discomfort Scale was used to assess frustration intolerance of an individual (Harrington, 2005). The construct describes how well an individual copes with frustration. Since frustration has been linked with video-telephony quality (Powers et al., 2011; Mrvelj & Matulin, 2021), general susceptibility to being frustrated is an interesting predisposition in the context at hand. The questionnaire differentiates between achievement, discomfort intolerance, emotional intolerance, entitlement, fairness, and also derives an overall score.

4.5.3 Personal Attributes

The term demography usually refers to an aggregation of variables to describe a sample or a population. This class of variables are used to describe individuals using such factors and is therefore referred to as *personal attributes*. For example: age, mother tongue (non-native speakers might evaluate conversations differently), if a person smokes regularly, begin an enrolled student (this variable was mostly assessed to keep track of how many people we recruited from outside the university), if a person has at least one academic degree, and the average time spent in video calls per week. In the same direction, it was also asked how important the person thinks video-telephony is in general, if they believe it would replace face-to-face communication in the future, and what their overall experience is regarding the average call quality of video-telephony calls.

Regarding the sexuality of the participant, two questions from the Genderbread Person were borrowed; it is not a questionnaire in a traditional sense, but rather an idea to think about different aspects of gender identity, such as anatomical sex or romantic attraction (Killermann, 2018). It offers two scales for each of the six dimensions, a feminine scale and a separate masculine scale. For example, the dimension "romantically attracted to..." comes with one scale to indicate the degree to which a person is romantically attracted to "women a/o feminine a/o female people" and an independent, second scale to which degree the person is romantically attracted to "men a/o masculine a/o male people", as both can be different than zero at the same time. In the context of this work, I focus exclusively on the romantic attraction dimension and the gender identification of the participants. It could be imagined that experiences of conversations vary, depending on if a conversation is with a person that one feels potentially romantically attracted to or not. Hence, this was added as a potential predictor.

4.6 Physiological and Behavioral Reactions

To assess the reactions a person shows when confronted with a certain stimulus, many different variables can potentially be obtained and investigated. While the literature provides plenty of examples of how to incorporate physiological measurements to distinguish a small number of different outcomes, such as in binary choice scenarios, there is not a lot of work concerned about estimating actual, continuous ratings. Especially, when looking at video-telephony, the body of literature is quite scarce. However, EEG, HR and HRV, EMG, eye tracking (including pupillometry), and EDA seem to be frequently used sensors that also have a lot of related work in their background (most important aspects are discussed in Sect. 2.2.4), attesting them good correlates with cognitive processes in the domain of sensory perception and experiencing different emotional states. This is the basis of the sensor selection in the rest of this work. However, EEG and EMG were neglected in the following chapters for practical reasons, as I opted to test a multitude of different sensors on the one hand and incorporate a meaningful number of trials and participants overall, I therefore wanted to minimize the setup time.

In Chap. 2, I described a subset of all these potential variables and compiled a set of relevant variables for all of the four components of the IAE model of experience that depend solely on the individual. However, some of these variables or measures allow to compute a sheer indefinite amount of features. For example, each of the sensory-based measures, sampled up to 500Hz, returns a large vector per trial that needs to be interpreted in order to be correlated with the defined outcome measures (scalars). This interpretation is a pre-processing step that converts a vector of many readings into scalar features. However, the literature knows many different, potentially useful features to derive from each sensor. Raw eye tracking data, for example, are often analyzed to count the number of fixations, or to compute an average duration of all fixations in an interval of interest. A second, frequently used metric is that describing eyelid blinks. Also pupillometry measures, such as the average pupil dilation or its variability, are best practices in the related literature.

This should demonstrate how a single sensor can produce a multitude of different features. In the case of the facial expression analysis described in Sect. 4.6.4, this vector to scalar conversion yielded 520 different scalar features. To mitigate a feature inflation, a factor analysis is performed on such large feature sets. However, for all the sensor-based features, first all highly correlating features are removed, respectively, only one of a set of highly correlating features is kept. Highly correlating means $r_{pearson} > 0.99$. Such features do not contain any additional information and can be ignored. In the case of the facial configuration analyses, a factor analysis is performed, to further reduce the number of features.

Since the two study paradigms have different stimulus durations, it is important to explain how the sensor-based features were computed: In the simulation experiment, participants watched short video clips of about 10 seconds duration. Here, the sensor features were computed on the data sampled over the entire duration of each trial. For the conversation experiments, all sensor-based features are computed taking the last 2 minutes of each conversation into account. Each conversation could be potentially up to 5 minutes, but since the participants had the option to terminate a call earlier, a 2-minute interval was chosen to have sufficient date for most of the trials.

4.6.1 ECG and HRV

A single electrode ECG was recorded using a Polar H10 chest strap. The wearable device records a 125 Hz ECG and provides either the entire raw data or pre-processed RRI; the time intervals between two subsequent R-peaks of the PQRST-complex of an ECG, colloquially referred to as the beat-to-beat interval. These RRI were used to derive min, max, and mean heart rate, and also to derive various HRV metrics. The literature offers many possible HRV metrics to consider. I selected a broad spectrum of different most cited HRV metrics, to be tested for relationships with the other related factors of the IAE model. These features are distinguishable in three categories:

- **Time domain features:** Min, max, average, and standard deviation of the heart rate; Root Mean Square of Successive RR-Interval Differences (RMSSD); and the *pNN50*-metric, which refers to the percentage of successive RRI that differ by more than 50ms. While the standard statistics are commonly used to describe physiological arousal and excitation states, the RMSSD and the pNN50 have been shown to decrease during stress significantly (Kim et al., 2018), and hence are an interesting choice for potentially frustrating and arousing audiovisual communication experiences.
- **Frequency domain features:** The low-frequency spectral power (LF) and the high-frequency spectral power (HF) play the most important role. Interestingly, it was observed that the HF band is influenced only by parasympathetic activity, while the LF band is influenced by both sympathetic and parasympathetic activity. Studies have shown that stress causes HF to decrease and LF to increase. Consequently, the ratio of LF to HF power (LF/HF) must increase. However, it has been shown that the ratio is a more accurate predictor than any of the parameters individually (Giannakakis et al., 2019). Nevertheless, all three features are included into the analysis.
- **Geometric features**: This class of features derives information from geometrical properties of, for example, the histogram of RRI of a time frame of interest. Of this class, only the *triangular index* was used. The triangular index gets its name from fitting a triangle to the height of each interval. The ratio of the total number of beats divided by the number of beats in the modal bin is called triangular index. The beats are grouped into bins, and the modal bin refers to the bin that contains the highest number of samples. The triangular index is essentially the area of the triangle divided by the area of the modal bin (Kleiger et al., 2005). The metric also shows promising predictive attributes to estimate stress and mental load (Ushiyama et al., 2008) and even emotional states (Rumpa et al., 2018).

ECG **based feature names**: `mean hr, min hr, max hr, std hr, rmssd, pNN50, triangular index, lf power, hf power, lf/hf ratio`

A HRV analysis requires at least a 1-minute interval of data to yield meaningful results. Since only the conversation experiment contains long enough data samples, the HRV analysis was not carried out for the simulation experiments, as their trails lasted only about 10 seconds on average.

4.6.2 EDA

The EDA signal was recorded using Shimmer GSR+ units. The wearable device is worn on the wrist of the non-dominant hand and exposed two electrodes that are strapped around the third segment of the index and the middle finger. The device samples the skin conductance signal at 51.2 Hz. For the processing of the raw signals, NeuroKit2 was employed (Makowski et al., 2021). The feature generation is oriented on the systematic analysis of different EDA features for emotional classification (Kołodziej et al., 2019). Therefore, the raw signal was split into its phasic and tonic components, before statistical values of the phasic SCR were derived as features:

- **Statistical description of the phasic component:** As suggested by Wagner et al. (2005), typical statistical descriptive values of the phasic component were shown to contain predictive information in an emotion recognition context. Hence, mean, median, standard deviation, variance, minimum, maximum, range, 95th percentile, and rmssd values were computed on the phasic samples.
- **SCR max peak**: Saffaryazdi et al. (2022) also used the following SCR features in an emotion recognition during conversations task: The maximum peak amplitude, the maximum peak rise time, the maximum peak skewness, as well as the maximum peak kurtosis. All these features describe different statistical properties of the same signal peak.
- **1s deltas and derivative features:** Kołodziej et al. (2019) further describes *maximum delta forward* and - *backward* features. Those are maximum value of the difference between the amplitudes of the local extrema and the amplitude values of the signal samples measured one second later or before the peak. They also derive two features regarding the sum of positive and negative values of the first derivative of the skin conductance signal.

While all these features are less intuitive to grasp, they all capture various statistical properties of the skin conductance response to what the participants experience. Further descriptions of the features are documented in Wagner et al. (2005) and Saffaryazdi et al. (2022). However, a correlation analysis among all the described features revealed very strong colinearities among elements of this feature set. Because the following analyses are mostly centered on linear models, such colinearites should be avoided and won't provide any additional, predictive value. Only one of the many highly correlated features was kept for the final set of features.

Altogether, four descriptive statistical features of the phasic signal, plus seven SCR based features were generated for the following analyses:

EDA-**based feature names**: `phasic min, phasic mean, phasic median, phasic rmssd, scr delta backwards, scr delta forwards, scr peak kurtosis, scr peak skew, scr peak rise time, scr peak count, scr sum neg differences`

4.6.3 Eye Tracking and Pupillometry

All eye-related data were recorded using a Tobii Pro Nano stationary eye tracker. It was mounted to the bottom of the screen, and a five-point calibration routine was performed prior to each experiment. The eye-tracker records gaze and pupillometry data at a temporal resolution of 60Hz. Three feature categories are considered: fixations, blinks, and pupillometry.

Fixations and blinks were extracted from the raw gaze data using the R library *saccades* (von der Malsburg, 2019). This yielded the number of *fixations per minute*, the *mean and the media duration of the fixations* as features, as well as the number of *blinks per minute*, and *mean and median duration of the blinks* per trial (only recordings of exact 2 minutes were taken into account).

Pupillometry has been shown effective in the classification of affective states (Zheng et al., 2020; Sirois & Brisson, 2014) and cognitive load (Hess & Polt, 1964). The stream of pupil diameters was analyzed regarding the *mean* dilation, the *5th percentile* and the *95th percentile*, and the *RMSSD* of the pupil dilation. Duchowski et al. (2018) introduce a new eye-tracked measure, the Index of Pupillary Activity (IPA), as an alternative to the proprietary Index of Cognitive Activity (ICA) for measuring cognitive load from pupillometry samples.

Altogether, six fixation and blink features, as well as five pupillometry-based features, were generated for the following analyses.

Eye-related feature names: `fixations per min, mean fixation duration, median fixation duration, blinks per min, mean blink duration, median blink duration, mean pupil dilation, pupil dilation 95%, pupil dilation 5%, pupil dilation rmssd, pupillary activity index`

4.6.4 Facial Configuration Tracking and Analysis

Facial configurations, the term that describes the analysis of facial movements and their changes over time (Barrett, 2017; Barrett et al., 2019), sometimes also called facial expressions, were recorded using a custom made iOS app for an Apple iPhone 13 Pro. Apple's ARKit library provides real-time measurements of 52 different facial landmarks, accessible via the FaceTracking API and the *blendShapes* property of the *ARFaceAnchor* object (Apple Inc, 2023). These landmarks were continuously monitored at approximately 35.6Hz, after dropping subsequent duplicates.

For the analysis of the data, only time intervals of stimulus exposure were considered. For the simulation study, those were all the time intervals, while participants watched the stimulus videos, and for the conversation experiments, those were all conversation intervals (as opposed to the instances of filling in questionnaires between the trials, for example). For each of these intervals and each of the 52 facial markers, we computed a multitude of different features:

- **Mean** value over the entire interval.
- **Standard deviation** over the entire interval.
- Root mean square of successive differences (**RMSSD**) as a measure of variability.
- **Activity ratio**; the number of samples having a value higher than the grand average of samples of the same feature, divided by the total number of samples in the given interval. It describes how much of the time this value was above average.
- **Frequency analyses** of the interval signal, based on a power spectral density estimation using a periodogram. To distinguish the low from the high-frequency bands, the threshold is set to be half the total power of the signal. The entire dataset has been analyzed to determine these thresholds for each dimension. The low-frequency (LF) band ranges from 0 to 0.5 * total_power of an average signal of a given dimension; the high-frequency (HF) band from 0.5 * total_power to total_power. This procedure is inspired by frequency analyses of Heart Rate Variability analyses (Spiers et al., 1993; Champseix et al., 2021). To reproduce this computation, see Spang et al. (2023b).
  - **LF band** power.
  - **HF band** power.
  - **Full band** or total power.
  - **Ratio** between LF and HF.
  - **Percentage of LF** in the total power.
  - **Percentage of HF** in the total power.

This feature generation process resulted in ten scalar features per dimension and a total of 520 features per trial. To find out which of our 520 features are meaningful, a factor analysis was performed, reducing the set to 14 factors. These 14 factors explained 75.828% of the variance of the dataset of the simulation studies, and 49.336% of the variance of the conversation experiment data. Prior to the factor analysis, features without any variability ($SD = 0$) were filtered out, as well as such with high colinearity to other features ($r > 0.99$; one was kept, the others were dropped). For the factors, only features with a high loading of at least 0.5 and only factors with at least three features were considered. This procedure and a discussion regarding the stability of the factors across experimental paradigms are further elaborated on in Spang et al. (2023b).

Facial Configuration Feature Names: `face factor 1..14`

4.6.5 Conversational Interactivity

As introduced in Sect. 2.2.4.5, Hammer, Reichl, and their colleagues proposed metrics to quantify interactivity between two conversational partners (Hammer & Reichl, 2005; Hammer et al., 2005; Reichl & Hammer, 2004). For the purposes of this research, their metrics have been adapted. Instead of solely focusing on the speaker characteristics of speaker A and B, data are gathered for each participant, incorporating their perspectives as both self and other. This added dimension enables the use of the proposed metrics in conjunction with other forms of data such as sensor-based measurements or subjective responses, aligning an individual's behavior with, for instance, their own questionnaire responses.

Based on their work, the following interactivity metrics are used to describe the engagement of a person in a conversation on the one hand, and, in the sense of the IAE model, use interactivity for the estimation of perceived QoE. Therefore, 16 different interactivity parameters are taken into account: the four state probabilities, individually from the perspective of each participant (self-talk and interlocutor talk, mutual scilence, and double talk), the respective sojourn times (sojourn time self, sojourn time interlocutor, sojourn time mutual silence, and sojourn time double talk), the speaker alternation rate, the interruption rate, the double talk rates for both conversation partners, the pause rates for both and the mean time of pauses between speaker changes, and the conversational temperature.

Conversational interactivity feature names: `prob mutual silence`, `prob self-talk`, `prob other talk`, `prob double talk`, `sojourn mutual silence`, `sojourn self`, `sojourn other`, `sojourn double talk`, `speaker alternation rate`, `interruption rate`, `pause rate self`, `pause rate other`, `mean pauses speaker change`, `double talk rate self`, `double talk rate other`, `conversational temp`

In the following experients, two paradigms are introduced; a passive viewing/listening experiment and an active conversational setting with two participants talking to each other. The intercativity analysis is only applicable to the conversational experiment, hence all features discussed are neglected when analyzing the passive experiment in Chap. 5.

Chapter 5
Relationship Validation

The definition of insanity is doing the same thing over and over again and expecting different outcomes.
Unknown author, often attributed to A. Einstein

Defying the sentiment of the opening epigraph, this chapter retests hypotheses to highlight consistent predictors with statistically significant results across tests. These predictors then inform future predictive modeling in subsequent chapters.

Divided into two parts, this chapter first employs a controlled simulation approach to examine the impact of quality degradation on responses, mimicking a video-call scenario. Findings from this simulation are then used to scrutinize each aspect of the IAE model through statistical testing. The second part follows a more naturalistic approach, examining quality degradation impacts on real conversations. Findings from both parts are compared to assess the reproducibility of the observed relationships.

This chapter serves dual purposes. Firstly, it offers evidence supporting the IAE model of experience introduced earlier (refer Sect. 3.3), without prematurely discussing the predictability aspect, which is reserved for Chap. 7. Secondly, it presents a unique framework for conducting video-based conversation simulations. The first half of this chapter leverages this simulation framework to provide preliminary insights into its potential utility. These findings will then be replicated in the second half, using a different experimental paradigm to assess their robustness.

Novelty here stems from applying existing methodologies to previously unexplored connections. The focus is not on the testing methods but on how physiological and behavioral measures can facilitate the study of individually perceived quality. Predictors with significant relationships to various outcome measures are enumerated, but their relative importance will be addressed in Chap. 6 using an SEM approach.

R. P. Spang, *Individualized Quality of Experience Estimation in Audiovisual Communication*, T-Labs Series in Telecommunication Services,
https://doi.org/10.1007/978-3-031-67551-5_5

To identify which features and variables display significant relationships, this chapter will carry out various statistical tests, adopting a breadth-first search approach dedicated to gathering evidence for the edges of the IAE model. Despite the necessity of considering all available data, the sheer volume of analyses presents an increased likelihood of Type-I errors. These errors, indicative of falsely attributing significance to a relationship due to a statistical anomaly, are a common concern when performing multiple statistical tests on the same dataset as discussed in the literature (Bland & Altman, 1995; Cabin & Mitchell, 2000). A common mitigation technique for such an issue is the application of an alpha-level correction such as the Bonferroni–Holm correction, as outlined by Holm (1979). However, this approach risks increasing the likelihood of Type-II errors. In light of these potential pitfalls and the overarching aim to identify all possible connections in a data-driven manner, a decision has been made to forego such a correction in this instance. To ensure the validity and reliability of the findings, any identified relationships will be retested in the latter half of this chapter. This retesting will utilize a different experimental paradigm, different data, and participants, thus only considering those relationships that persist across these validations as stable and relevant.

This methodology enables the testing of all possible relationships without making alpha-level corrections. The findings presented in Sect. 5.1.2 will include feature names and correlation results. However, given the sheer number of features within the reaction dimensions, only the most statistically significant results are detailed in the chapter's tables for the sake of readability. An Open Science Framework (OSF) Repository contains all result tables created in the following analyses.[1]

The primary findings of this chapter are reported through tables, which encompass a predictor variable, a response variable, and three critical statistical measures:

- **Standardized β-value:** This value signifies the magnitude and the directionality of the relationship between a predictor variable and the response variable, with the scales of the variables being duly accounted for. The computation of the standardized β value involves transforming both the predictor and response variables into their z-scores. This standardization process establishes a uniform scale with a mean of 0 and a standard deviation of 1. In effect, the standardized β value reflects the changes in the response variable expressed in standard deviation units. This enables an unambiguous comparison of the relative impact of different predictor variables on the response variable. A positive β value points toward a positive correlation between the predictor and response variables, and the inverse is also true.
- **Standard Error (SE):** This metric denotes the accuracy or variability associated with the estimate of a statistical figure in a given analysis. It essentially quantifies the average error or uncertainty anticipated while estimating the accurate

[1] https://osf.io/h3uzr

population parameter. A reduced SE corresponds to heightened precision in the estimation, and vice versa.

- **p-value:** The p-value signifies the likelihood of encountering a test statistic as extreme as, or more extreme than, the observed data, assuming the validity of the null hypothesis. A p-value smaller than the α-level of 0.05 provides strong evidence against the null hypothesis, implying the preference of an alternate hypothesis. Conversely, a larger p-value suggests weak evidence against the null hypothesis, indicating that the observed data align well with the null hypothesis. Importantly, a p-value does not directly gauge the practical significance or the magnitude of an effect; it rather evaluates the probability of observing the given data under the presumption of the null hypothesis.

5.1 Quality Evaluations of Simulated Conversations

This section contains the analyses of the conversation simulation paradigm and presents the results of the experiments. I discuss the experimental setup, procedure, and task, building upon the information introduced in Sect. 4.1.1. A manipulation check is conducted to ensure that the quality manipulation successfully influenced user ratings. The primary focus of this section is the analysis of the IAE model validation in Sect. 5.1.2.

5.1.1 Experiment and Methods

While technically producing one dataset, the study was carried out two times, not including two pre-studies that were conducted earlier to test the experimental paradigm and the technical implementation. Despite these precautions, two study intervals were necessary to record the final dataset. Hence, the study as described here was originally recorded between March and May 2022 and exactly replicated between May and June 2023. Because of the identical replication, the data from both study periods were pooled for the following analyses.

5.1.1.1 Participants

I recruited 53 participants via a university participant database of TU Berlin. 49.1% identified as female, 11.3% chose not to answer how they identify. The median age was 29 years (IQR: 11y, ranging from 21 to 58y). In total, 84.3% indicated German as their first language; the remaining spoke fluent German; 26 participants hold at least one academic degree; 51.0% were currently enrolled, students. Participants received 15 Euros per hour as monetary compensation; the experiment lasted 1 hour on average. The ethics committee approved the study protocol (ethics ID:

FR_2022_03) of Technical University Berlin, Faculty IV Electrical Engineering and Computer Science. The conductance of the experiment was according to the declaration of Helsinki. The participants obtained informed consent in written form and declared their agreement with the procedure by signature before the recordings began.

5.1.1.2 Evaluation of the Conversation Simulation

Regarding the **Subjective Quality Ratings**; participants rated the overall quality of the data set clips between "excellent" and "ideal" (median = 5.111, IQR = 1.071) in degradation level 1, between "fair" and "good" (median = 3.087, IQR = 1.5) in level 2, as "bad" (median = 1.008, IQR = 0.818) in level 3, and as "very bad" (median = 0.56, IQR = 0.859) in level 4. Friedman's test (Daniel, 1990) showed significant differences between the quality groups ($Q = 978.149$, $p < 0.001$). Post-hoc contrasts using Wilcoxon's Signed-Rank Test (Wilcoxon, 1945) show significant differences between all pairs of quality groups (contrasts resulted in significant differences of $p < 0.001$ after adjusting for multiple comparisons). Ratings for the video and speech quality separately, among further details of how the dataset was perceived, are discussed in Spang et al. (2022b). Because participants clearly distinguish all provided degradation levels of the clips, and because all four narratives provided were evaluated to be comparable regarding their likeability and vivacity, I conclude that the dataset is generally suitable for audiovisual conversation simulation. Figure 5.1, left side, provides a comparison of all three quality dimensions for each of the four degradation levels of the dataset.

Regarding the **Content Related Responses**; responses to the content-related questions were, on average 87.5% correctness for degradation level 1 (IQR = 25%),

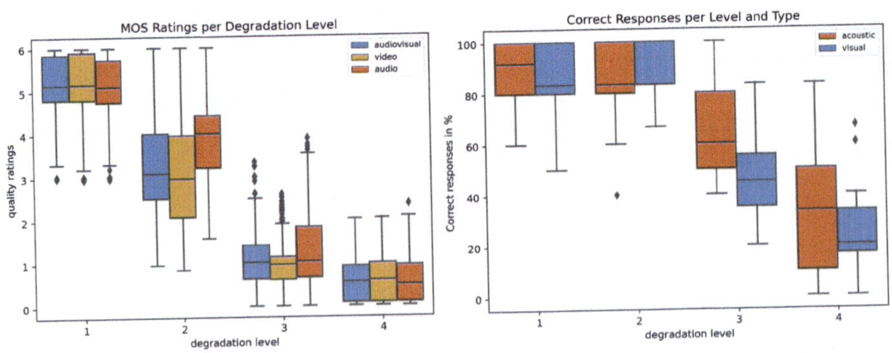

Fig. 5.1 Left: MOS comparison per degradation level (1 = no impairment, 2 = reduced quality, 3 = short dropouts, and 4 = mostly frozen). Ratings were administered through a seven-point MOS scale including overflow areas on both sides, scores range from 1 = "very bad" to 7 = "ideal". Right: Percentage of correct responses per degradation level and question type. After each trial, a question was asked regarding either the visual or acoustic content of the video

87.5% correctness for degradation level 2 (IQR = 25%), 57.143% correctness for level 3 (IQR = 16.667%), and 37.5% correctness for level 4 (IQR = 33.333%). The largest difference between responses to questions that referred to acoustic cues compared to questions regarding visual cues was observed in clips of degradation level 3. Here, a median difference of 16.667% was recorded: questions about what was said in the clips were answered correctly in 66.667% of the trials, whereas questions regarding what was shown were answered correctly in only 50% of the trials.

Friedman's test showed significant differences of the response correctness between the quality groups ($Q = 337.653$, $p < 0.001$). Post-hoc contrasts show no significant difference between the response correctness of level 1 and level 2 trials. However, the difference between level 2 and level 3 ($T = 44.0$, $p < 0.001$), and between level 3 and level 4 ($T = 30.0$, $p < 0.001$) was statistically significant. The difference between the question types (acoustic vs. visual) was not significant for any of the four levels. Figure 5.1, right side, compares the correct responses contrasting the four degradation levels of the database.

5.1.1.3 Systematic Analysis of IAE Model Relationships

This chapter expounded the reasoning behind the video-based simulation and furnished evidence supporting its effectiveness in delivering content across the four distinct degradation levels discussed in Sect. 4.1. This validation enables us to further explore subtle physiological and behavioral changes in participants using the described video clips within the simulation paradigm.

Section 5.1.2 that follows delves into a comprehensive analysis of the relationships within the IAE model, drawing upon the aforementioned dataset. Thus far, the discussion has primarily centered around the quality impressions as rated by the 53 participants post each trial. Moving forward, the focus will shift to other recorded data, encompassing elements like questionnaire responses about participants' personalities, and physiological responses registered during the video viewing. This broadened lens allows to identify general influencing factors and to address the research questions related to the video-based conversation simulation approach.

5.1.2 *Observations*

The subsequent subsections embody four independent yet integrated perspectives, each dissecting the IAE model utilizing conversation-simulation data:

1. Influence of Quality Degradation on Experience Metrics, States, and Reactions
2. Impact of Individuality on Experience Metrics, States, and Reactions

3. Role of Preceding States, Reactions, and Ratings on Subsequent Experience Metrics
4. Reactions as Indicators for Experience Metrics

The model captures and highlights an array of unique interconnections among its components. Here, a *perspective* denotes the dynamic pathway within the model that begins from an external element—signified by an outgoing arrow, that facilitates the construction of an experience, and terminates at another external element, signaled by an incoming arrow, that arises from the established experience. For instance, this may embody a progression from a specific stimulus quality to the corresponding reactions, or stem from an individualistic trait to physiological responses. Each perspective undergoes rigorous investigation via experimental data to validate the proposed relationships within the model.

In subsequent analyses, linear mixed effect models have been deployed to statistically validate each relationship suggested by the model. Grouped by the four perspectives delineating the flow of influences proposed by the IAE model, each statistical test represents a linear model of the form as depicted in Eq. 5.1 (in R-formula notation). If feasible, this models a random intercept for the predictor variable per participant, accommodating for varying effects of the predictor per participant. For instance, the model indicates that the stimulus's inherent quality affects the user's audiovisual quality rating. Here, some participants may rate the quality of the same video clip slightly higher than others. The random intercept on the participant level accounts for such interpersonal variations, yielding a general assessment of the relationship between a clip's quality and its user ratings. However, if the linear model cannot be fitted—potentially due to insufficient data available to validate a hypothesized relationship—a fallback analysis was performed without the random intercept, as a simple linear regression model.

$$\text{outcome} \sim \text{predictor} + (1 + \text{predictor}|\text{userId}) \tag{5.1}$$

The dataset obtained from the conversation simulation paradigm comprises a total of 1,575 trials involving 53 participants. It is important to note that not all trials have complete records of all intended variables, resulting in variations in sample sizes. This variability is due to factors such as sensor failure, network issues causing data loss in the lab, or participants not responding to all questionnaires. For the subsequent analyses, however, the maximum number of available samples per test was utilized.

Among the combinations, the one with the smallest number of samples consists of 110 samples from 20 unique participants. Despite this, maintaining maximum statistical power was a primary objective throughout the data collection process. As a result, even this particular pairing, which represents the least robust link, includes data from a minimum of 20 participants.

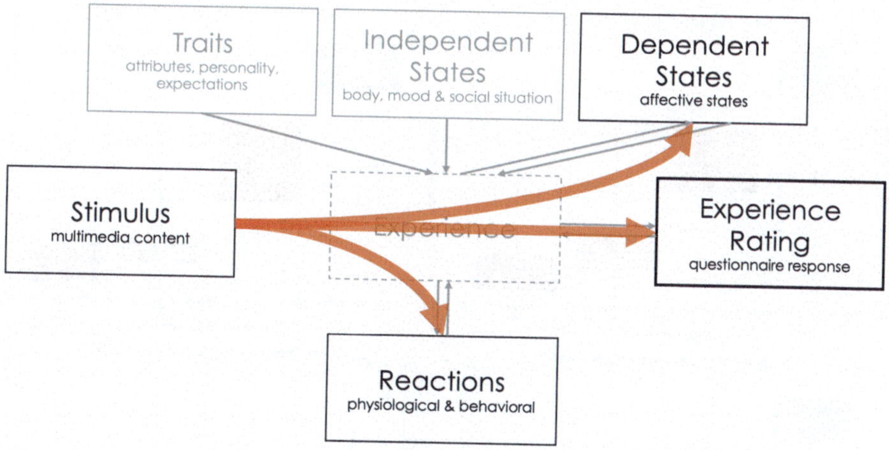

Fig. 5.2 Pathways of the IAE model showcasing how quality degradation impact *experience metrics*, *dependent states*, and *reactions*. This hypothesis is the focus of the current subsection. Compare the full model in Sect. 3.3

5.1.2.1 Quality Degradation Influences Experience Metrics, States, and Reactions

The initial perspective of this section aims to examine the direct influences of stimulus quality on all components of the IAE model that may be affected by changing experiences. Figure 5.2 illustrates the respective pathways. In this analysis, the predictor variable considered is the degradation level of the stimuli. The tests conducted focus on understanding the impact of stimulus quality on various elements of the model, including (experience) dependent states, the QoE outcome measures defined for this study, as well as physiological and behavioral reactions. A total of 49 relationships were examined, of which 23 relationships (46.94%) were found to be statistically significant.

The four levels of stimulus quality significantly influence QoE outcome variables, corroborating the relationship between the *Stimulus*, the latent variable *Experience*, and the *User Ratings*, particularly on the specified QoE metrics of interest. Figure 5.3 displays an overview of all significantly influenced outcome variables, while Table 5.1 offers additional statistical details about the relationships. The table also enumerates all significant correlations between the stimulus's degradation level, acting as a predictor variable, and all potentially influenceable dimensions. The β values denote standardized regression coefficients. All dependent states were impacted by the clips' quality—a negative sign indicates that as the quality worsens, participants experience lower valence and arousal and are more likely to terminate the call and redial. It should be noted that the degradation level is coded such that degradation level one represents the best quality, and degradation level four represents the worst, hence the interpretation of the negatively signed beta coefficient.

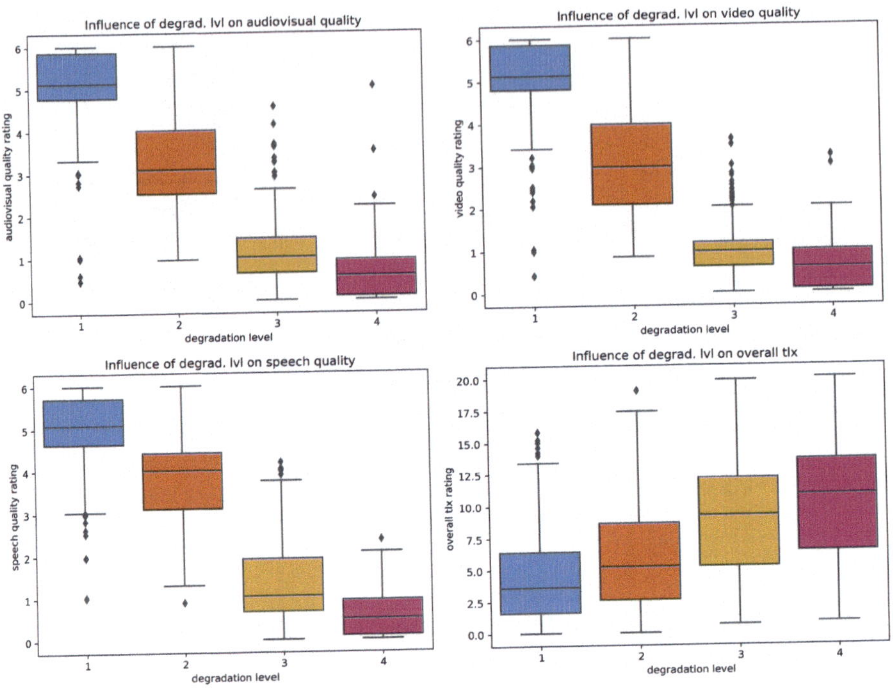

Fig. 5.3 Validation of manipulation on outcome measures: the degradation levels of the stimulus clips exert statistically significant influences on all outcome dimensions. Degradation level 1 corresponds to the best quality, 4 to the worst

With regard to outcome measures (referring to Sect. 4.2), all three quality ratings negatively correlate with the stimuli's degradation level, implying that the quality ratings diminish as the stimuli's quality deteriorates. In contrast, all TLX items are positively correlated with the degradation level. This suggests that participants experienced a higher task load with low-quality clips compared to high-quality ones (see Fig. 5.3).

Lastly, the β values for the reaction class variables—considered here as outcome variables—indicate weaker relationships, even though all three classes of measurements, namely facial configurations, eye tracking, and EDA, are represented with statistical significance. The sensor-based relationships are less intuitive to interpret. Generally, a positive β coefficient implies *more activity* as the stimulus quality declines, though this activity is contingent on the sensor and its feature. A comprehensive description of each relationship will be addressed in Sect. 5.3 while discussing the final set of features.

Table 5.1 Presented further is a compilation of significant correlations between the stimuli's degradation level and three variable classes directly influenced by the stimuli. Non-significant relationships are excluded. The table is ordered by outcome type and absolute beta, with the strongest correlations placed at the top. All dependent state variables are significantly influenced by the degradation level, as are all QoE variables. Moreover, the degradation level also impacts various reaction metrics, encompassing all different types of reaction measurements

Predictor	Outcome type	Outcome	β	SE	p
Degrad. lvl	Dependent state	Preference to abandon	−0.864	0.008	<0.001
		Valence	−0.811	0.004	<0.001
		Arousal	−0.309	0.011	<0.001
Degrad. lvl	Outcome measure	Speech quality	−0.895	0.041	<0.001
		Audiovisual quality	−0.888	0.046	<0.001
		Video quality	−0.885	0.041	<0.001
		Performance	0.508	0.229	<0.001
		Overall TLX	0.481	0.185	<0.001
		Frustration	0.442	0.245	<0.001
		Effort	0.427	0.254	<0.001
		Mental demand	0.386	0.318	<0.001
		Temporal demand	0.084	0.181	0.038
Degrad. lvl	Reaction	Face factor 14	0.238	0.010	<0.001
		Face factor 7	0.210	0.038	<0.001
		Scr peak kurtosis	0.161	0.055	0.002
		Fixations per min	0.141	0.003	0.033
		Mean blink duration	0.123	1.549	0.021
		Face factor 12	0.118	0.020	0.010
		Face factor 8	−0.086	0.066	0.013
		Face factor 4	0.068	0.051	0.048
		Pupil dilation 5%	0.016	0.066	0.013
		Pupil dilation 95%	0.012	0.047	0.002

5.1.2.2 Individuality Influences Experience Metrics, States, and Reactions

This subsection examines the influence of individual psychological aspects—*traits* and *independent states*—on *dependent states*, defined QoE outcome metrics, and *reactions*. Figure 5.4 depicts the pathways relevant to the analyses, which encompass a total of 1,323 relationships. Of these, 150 or 11.34% were found to be statistically significant.

All tested influences, which were administered only once and thus remain static throughout the experiment's trials, pertain to the relationships explored. Meanwhile, the outcome variables derived from the following analyses were recorded for each trial. A significant relationship, therefore, denotes a systematic difference between participants with certain attributes, regardless of any stimuli or quality experience.

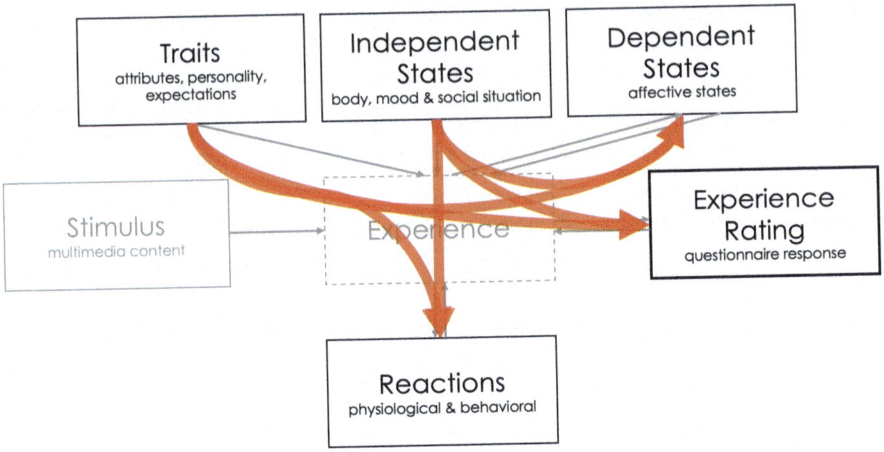

Fig. 5.4 IAE model pathways probing the impact of psychological individuality aspects. Tests are conducted to investigate the influence of traits and individual states on dependent states, defined QoE outcome metrics, and individuals' reactions

Table 5.2 enumerates all statistically significant personality dimensions influencing terminal metrics. All five personality dimensions significantly correspond with at least one other metric. Notably, agreeableness influences only speech quality ratings, while extraversion and openness affect a wider range of variables, eight or nine, respectively.

Table 5.3 addresses the frustration discomfort dimensions, also part of the collected traits. Like the personality traits, the Frustration Discomfort Scale (FDS) was administered only once. Both tables thus highlight static influences on other states, reactions, and QoE metrics. Interestingly, frustration intolerance metrics do not appear to influence any dependent states, such as valence or arousal. However, they do seem to impact facial configurations, eye-tracking related variables, and effort and task load-related ratings.

The final component of the traits analyses focuses on personal attributes, typically considered demographic factors. Here, they are evaluated as predictors of other model variables and aspects, hence the term "personal attributes," to differentiate them from aggregated demographic data. With 71 significant relationships, Table 5.4 only includes those with a standardized beta greater than 0.2 for readability. The complete list is available at https://osf.io/h3uzr. All attributes significantly influence other variables, particularly reaction-class variables.

Finally, the independent state variables are evaluated within the IAE model framework, as presented in Table 5.5. While mostly reaction variables show significant relationships with sleep and hunger-related states, the time since participants last ate something displays a direct link to their quality response behavior. This phenomenon, sometimes known as "hangriness," was the subject of a more in-depth investigation in Spang et al. (2023c).

Table 5.2 Table listing significant personality dimensions—a subset of the recorded trait variables—and their impact on dependent states, QoE experience ratings, and reactions

Predictor	Outcome type	Outcome	β	SD	p
Agreeableness	Outcome measure	Speech quality	−0.036	0.117	0.024
Conscientiousness	Reaction	Fixations per min	−0.347	0.028	0.026
		Mean fixation duration	0.186	84.162	0.029
		Phasic mean	−0.183	0.011	0.011
Extraversion	Outcome measure	Temporal demand	0.248	2.180	0.049
		Effort	0.147	1.147	0.009
	Reaction	Face factor 11	−0.340	0.471	0.018
		Scr sum neg differences	0.246	0.260	0.039
		Phasic min	0.230	0.081	0.045
		Mean pupil dilation	0.203	2.093	<0.001
		Face factor 10	−0.155	0.064	0.028
Neuroticism	Outcome measure	Temporal demand	0.215	1.302	0.037
	Reaction	Median blink duration	−0.330	11.623	0.050
		Mean blink duration	−0.316	12.780	0.049
Openness	Dependent state	Valence	0.082	0.031	0.009
	Outcome measure	Effort	0.190	1.474	0.012
		Mental demand	0.171	1.503	0.032
	Reaction	Face factor 5	−0.354	1.120	0.045
		Face factor 11	−0.340	0.639	0.036
		Face factor 6	−0.333	0.569	0.014
		Scr delta forwards	−0.261	0.060	0.034
		Phasic min	0.260	0.098	0.025
		Scr peak rise time	−0.251	0.645	0.020

5.1.2.3 Preceding States, Reactions, and Ratings Influence Subsequent Experience Metrics

This subsection investigates the impact of pre-existing or prior states, ratings, and responses on QoE measures, independent of any stimulus quality. Essentially, it seeks to discern which metrics are influenced, and to what extent, by a person's state or response when confronted with the next stimulus. This information is vital as it sheds light on the inertia or persistence of opinions. For example, if a person has a strong inclination toward a particular state, this could influence their subsequent ratings. With the dataset from the experiment, I can probe this relationship not only for states and experience ratings but also for physiological responses. Consequently, the forthcoming analyses will present correlation results between previous states and responses and subsequent stimuli.

To execute this analysis, data from each participant's trial A were aligned with the subsequent trial B. For trial B, I assessed the relationship between previous responses (from trial A) and the current experience ratings. It is important to

Table 5.3 Table listing significant frustration intolerance dimensions—a subset of the recorded trait variables—and their impact on dependent states, QoE experience ratings, and reactions

Predictor	Outcome type	Outcome	β	SE	p
Achievement	Outcome measure	Overall TLX	0.175	0.549	0.021
		Correct response	−0.075	0.019	0.004
	Reaction	Fixations per min	−0.426	0.010	0.003
		Mean fixation duration	0.264	42.680	0.031
		Median fixation duration	0.242	27.068	0.027
Discomfort intolerance	Outcome measure	Temporal demand	0.289	0.877	0.003
		Overall TLX	0.189	0.760	0.033
		Performance	0.109	0.577	0.017
	Reaction	Face factor 13	−0.349	0.156	0.005
		Face factor 9	0.269	0.283	0.040
		Face factor 12	−0.231	0.089	0.034
		Face factor 14	0.134	0.019	0.011
Emotional intolerance	Outcome measure	Frustration	0.165	0.618	0.039
	Reaction	Median fixation duration	0.270	23.776	0.013
		Mean fixation duration	0.263	39.245	0.038
		Face factor 12	−0.238	0.080	0.034
Entitlement	Reaction	Face factor 9	0.305	0.257	0.046
		Phasic mean	−0.066	0.001	0.025
Fairness	Outcome measure	Effort	0.188	0.715	0.022
		Mental demand	0.185	0.695	0.024
		Performance	0.089	0.461	0.048
	Reaction	Face factor 9	0.315	0.241	0.023
		Face factor 1	0.171	0.132	<0.001
Fds score	Outcome measure	Performance	0.093	0.618	0.039
		Correct response	−0.066	0.024	0.011
	Reaction	Face factor 13	−0.361	0.242	0.046
		Face factor 9	0.315	0.355	0.038
		Face factor 12	−0.250	0.103	0.032
		Face factor 14	0.183	0.015	<0.001
		Face factor 3	0.133	0.062	0.003

note that the conversation simulation study was designed such that a degradation condition always spanned two successive trials. As illustrated in previous subsections, a stimulus significantly influences many aspects, most notably the experience measures. Therefore, for every second trial, the preceding trial maintained the same degradation level. If the previous ratings were examined for inertia or carryover effects, this would merely reiterate the analyses already discussed earlier. To avoid any potential bias from a stimulus of one degradation level affecting the next stimulus of the same degradation level, this analysis only included trials preceded

Table 5.4 Table listing significant personal attributes—a subset of the recorded trait variables—and their impact on dependent states, QoE experience ratings, and reactions. To enhance readability, only relationships with a beta absolute value of at least 0.2 are included

Predictor	Outcome type	Outcome	β	SE	p
Age	Reaction	Face factor 8	0.527	0.027	<0.001
		Face factor 6	0.389	0.031	0.047
		Face factor 5	0.385	0.037	0.016
		Pupil dilation rmssd	0.381	0.007	0.002
		Median blink duration	0.248	0.406	0.017
		Face factor 7	0.225	0.013	0.020
Call time per week	Reaction	Pupil dilation rmssd	0.429	0.022	0.006
		Fixations per min	0.356	0.002	0.004
		Face factor 12	0.309	0.024	0.025
		Face factor 5	0.273	0.051	<0.001
		Face factor 11	−0.220	0.039	0.035
		Median fixation duration	−0.209	4.822	0.009
		Mean blink duration	0.200	1.103	0.036
Education class	Reaction	Mean fixation duration	−0.304	57.264	0.008
		Face factor 11	−0.299	0.328	0.042
		Face factor 9	0.281	0.373	0.043
gb rom. attract m	Reaction	Scr delta forwards	−0.333	0.008	0.004
		Scr sum neg differences	0.326	0.042	0.005
		Phasic min	0.310	0.013	0.007
		Scr delta backwards	−0.276	0.012	0.017
		Phasic mean	−0.222	0.001	0.007
Gender	Reaction	Scr sum neg differences	0.443	0.120	<0.001
		Scr delta forwards	−0.409	0.024	0.001
		Phasic min	0.387	0.042	0.003
		Phasic mean	−0.311	0.006	0.002
		Scr delta backwards	−0.310	0.036	0.007
		Scr peak count	−0.300	0.263	0.037
		Face factor 7	0.262	0.263	0.034
		Scr peak rise time	−0.236	0.333	0.023
		Face factor 14	0.224	0.041	0.012
Gratification	Reaction	Face factor 11	0.385	0.198	0.014
		Face factor 9	0.367	0.245	0.015
Handedness	Reaction	Face factor 5	−0.261	0.323	<0.001
Mother tongue	Dependent state	Arousal	−0.210	0.045	0.007
	Reaction	Face factor 11	0.353	0.391	0.008
Occupation class	Reaction	Face factor 7	−0.294	0.246	0.011
Romantic attract m	Reaction	Scr delta forwards	−0.246	0.009	0.040
		Face factor 7	−0.357	0.067	0.003
		Face factor 14	−0.304	0.007	<0.001
		Face factor 13	−0.269	0.054	0.043

Table 5.5 Table listing significant independent state variables and their influence on dependent states, QoE experience ratings, and reactions

Predictor	Outcome type	Outcome	β	SE	p
Being hungry	Reaction	Face factor 1	−0.132	0.827	0.044
		Median fixation duration	−0.015	−0.000	<0.001
		Blinks per min	−0.013	−0.000	<0.001
Being tired	Dependent state	Arousal	−0.182	0.054	0.008
	Reaction	Pupil dilation 95%	−0.704	6.277	<0.001
		Mean pupil dilation	−0.703	5.829	<0.001
		Pupil dilation 5%	−0.698	5.357	<0.001
		Phasic rmssd	0.499	0.001	<0.001
		Mean blink duration	−0.385	13.470	0.005
		Blinks per min	0.368	0.592	<0.001
		Median blink duration	−0.356	12.554	0.015
		Phasic min	−0.339	0.076	0.014
		Pupil dilation rmssd	−0.272	0.078	<0.001
Food abstention	Dependent state	Preference to abandon	−0.077	0.007	0.009
		Valence	−0.077	0.007	0.043
	Outcome measure	Audiovisual quality	−0.069	0.034	0.017
		Video quality	−0.061	0.034	0.035
	Reaction	Face factor 4	0.287	0.126	0.026
		Face factor 5	0.266	0.050	<0.001
		Scr peak skew	0.095	0.019	0.025
Hours of sleep	Reaction	Face factor 11	0.275	0.153	0.036
		Face factor 7	0.224	0.120	0.042
		Scr peak rise time	−0.203	0.164	0.033
		Pupillary activity index	0.191	0.011	0.032
		Pupil dilation 95%	−0.141	0.622	0.001
		Mean pupil dilation	−0.108	0.581	0.012

by trials of a different degradation level. This approach ensures that only genuinely different quality sequences are examined for inertia or carryover effects.

Figure 5.5 illustrates the influences being explored in this subsection. A total of 490 relationships were tested, and 67 (13.67%) were found to be significant. The data used in the subsequent analyses were collected after each trial of the experiment.

Table 5.6 enumerates the *dependent state* and *reaction* variables that significantly influence subsequent QoE outcome measures in the following trial. The statistically significant influence of each previous *reaction* type and *dependent state* variable on outcome ratings, independent of the stimuli themselves, validate the feedback pathway in the IAE model leading back to the *Experience* component.

Table 5.7 presents the significant relationships between ratings and subsequent ratings in the next trial. Similarly, the statistical evidence underpins the feedback loop in the IAE model from the experience ratings back to the latent experience

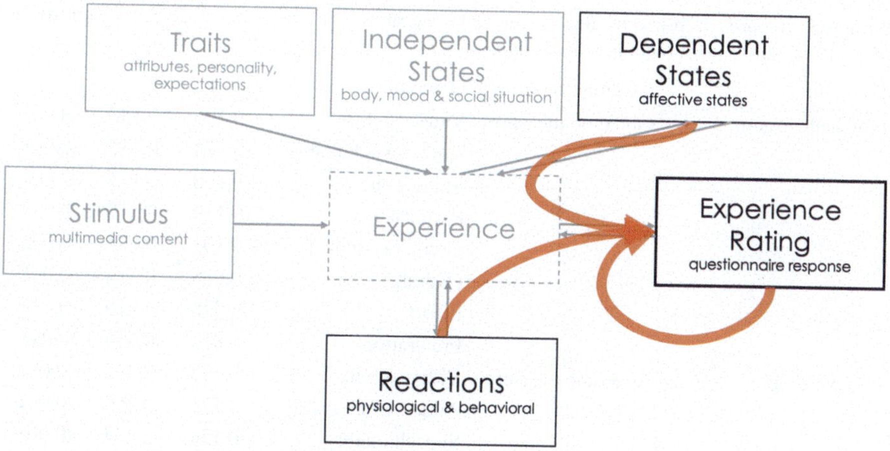

Fig. 5.5 The emphasized pathways in the IAE model illustrate the impacts of preceding dependent states, experience ratings, and reactions on ensuing QoE ratings

element, as expounded in Sect. 3.3.4.2. Notably, quality-related ratings influence nearly the entire gamut of QoE dimensions defined in this study, while effort and task load-related variables impact only a handful of variables, most of which are also quality-related.

5.1.2.4 Reactions as a Proxy for Experience Metrics

Lastly, in this series of model pathway examinations, I explore a relationship that, while not explicitly outlined in the IAE model of experience, is pivotal to this entire research project: the correlation between the reactions and the QoE outcome measures. This represents a fundamental aspect of the theoretical model; it is only when the observable reactions provide insight into a person's experiences that they can contribute effectively to the final estimation model. However, the correlation between reaction variables and user ratings, as illustrated in Fig. 5.6, is not a conventional pathway of the model. This is because, despite the necessary correlation, reactions are never directly correlated in practical applications per the ethos of the IAE model of experience. Instead, they are integrated into a estimation model that consolidates various signal sources, including user-specific static knowledge and multimedia signal information, along with reactions. This estimation model then calculates an individualized QoE metric score, which, owing to its multitude of influence variables, should be stronger than any of its inputs. This topic will be addressed in the forthcoming Chap. 7. Please note that the arrow from *Reactions* to *Experience Rating* in Fig. 5.5 represents the feedback loop from previous trials. On the other hand, the blue arrow in Fig. 5.6 illustrates the direct correlation between the *Reactions* and *Experience Ratings* within the same trial.

Table 5.6 Record of previous dependent state variables and reactions that exert significant influence on subsequent outcome dimensions

Predictor	Outcome type	Outcome	β	SE	p
Prev tech. accept.	Outcome measure	Video quality	−0.196	0.180	<0.001
		Audiovisual quality	−0.184	0.168	<0.001
		Speech quality	−0.170	0.175	<0.001
		Performance	0.110	0.515	0.002
		Correct response	−0.101	0.034	0.006
		Overall TLX	0.085	0.316	0.005
		Effort	0.078	0.410	0.010
		Frustration	0.070	0.456	0.048
Prev valence	Outcome measure	Video quality	−0.192	0.295	<0.001
		Audiovisual quality	−0.172	0.256	<0.001
		Speech quality	−0.154	0.237	<0.001
		Correct response	−0.100	0.048	0.008
		Performance	0.092	0.720	0.011
		Overall TLX	0.079	0.475	0.015
		Effort	0.074	0.634	0.031
Prev face factor 6	Outcome measure	Performance	0.246	0.448	0.003
		Video quality	0.163	0.153	0.030
Prev mean fixation dur.	Outcome measure	Correct response	−0.157	0.000	0.041
Prev fixations per min	Outcome measure	Correct response	0.199	0.758	0.009
Prev phasic min	Outcome measure	Performance	0.201	2.227	0.011
		Effort	0.148	1.236	0.004
		Mental demand	0.140	1.292	0.010
Prev phasic rmssd	Outcome measure	Mental demand	−0.148	89.850	<0.001
		Effort	−0.098	99.761	0.037
Prev scr delta forwards	Outcome measure	Mental demand	−0.195	1.637	<0.001
Prev scr sum neg diff.	Outcome measure	Performance	0.177	0.737	0.031
		Mental demand	0.126	0.396	0.016

In total, 360 relationships were tested, 47 of which were significant, accounting for 13.06%. Table 5.8 displays these significant relationships between the reaction dimensions and relevant QoE outcome measures, filtering for a standardized beta greater than 0.2 for better readability. The complete list can be found at https://osf.io/h3uzr. Most notably, all signal types used in the simulation study, including eye tracking, facial configuration tracking, and EDA, are represented in the list of significant relationships.

Table 5.7 Compilation of QoE ratings that significantly influence subsequent QoE ratings

Predictor	Outcome type	Outcome	β	SE	p
Prev audiovisual quality	Outcome measure	Video quality	−0.201	0.036	<0.001
		Audiovisual quality	−0.177	0.038	<0.001
		Speech quality	−0.160	0.037	<0.001
		Correct response	−0.093	0.007	0.012
		Performance	0.076	0.110	0.031
		Overall TLX	0.062	0.072	0.048
Prev speech quality	Outcome measure	Video quality	−0.220	0.036	<0.001
		Audiovisual quality	−0.200	0.037	<0.001
		Speech quality	−0.184	0.038	<0.001
		Correct response	−0.119	0.007	0.001
		Performance	0.096	0.108	0.007
		Overall TLX	0.077	0.066	0.008
		Effort	0.072	0.090	0.022
Prev video quality	Outcome measure	Video quality	−0.202	0.038	<0.001
		Audiovisual quality	−0.182	0.036	<0.001
		Speech quality	−0.162	0.037	<0.001
		Correct response	−0.088	0.007	0.016
		Performance	0.076	0.109	0.032
		Overall TLX	0.061	0.066	0.037
Prev correct response	Outcome measure	Video quality	−0.110	0.185	0.003
		Audiovisual quality	−0.096	0.205	0.019
Prev effort	Outcome measure	Video quality	0.099	0.013	0.007
		Audiovisual quality	0.092	0.013	0.013
		Speech quality	0.086	0.013	0.019
Prev frustration	Outcome measure	Video quality	0.127	0.013	<0.001
		Audiovisual quality	0.120	0.013	0.001
		Speech quality	0.111	0.016	0.008
Prev mental demand	Outcome measure	Audiovisual quality	0.099	0.014	0.015
		Video quality	0.098	0.017	0.037
		Speech quality	0.088	0.013	0.017
Prev performance	Outcome measure	Video quality	0.138	0.013	<0.001
		Audiovisual quality	0.130	0.012	<0.001
		Speech quality	0.120	0.012	0.001
Prev temporal demand	Outcome measure	Temporal demand	0.177	0.061	0.003
		Frustration	0.176	0.076	0.009
		Overall TLX	0.145	0.050	0.007
		Performance	0.144	0.063	0.004
Prev overall TLX	Outcome measure	Video quality	0.126	0.016	<0.001
		Speech quality	0.120	0.025	0.034
		Audiovisual quality	0.120	0.019	0.005

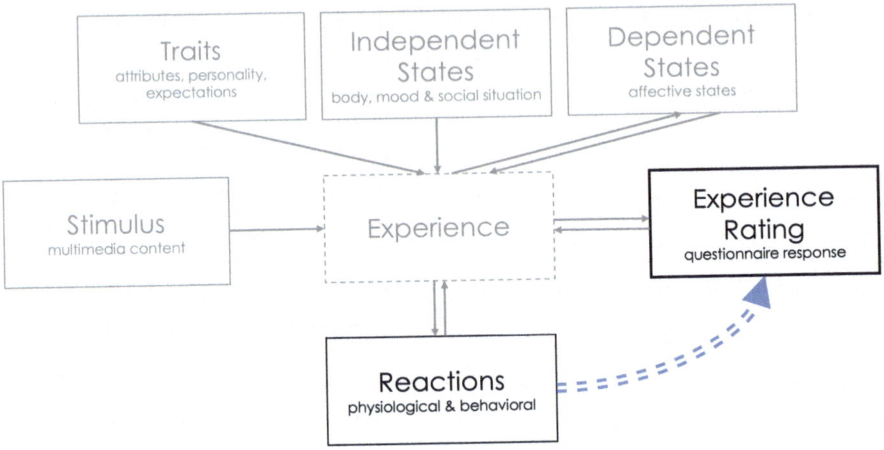

Fig. 5.6 The relationship between the reaction and user ratings entities in the IAE model. Although not explicitly illustrated in the model, this relationship forms the bedrock of how individual influences enhance QoE estimations. For a discussion about the components of the model, refer to Sect. 3.3

Table 5.8 Catalog of reaction variables recorded during the conversation simulation experiment and their significant influence on outcome dimensions

Predictor	Outcome type	Outcome	β	SE	p
Face factor 14	Outcome measure	Effort	0.279	2.507	0.006
		Mental demand	0.275	2.495	0.006
		Overall TLX	0.252	1.849	0.009
		Speech quality	−0.223	0.458	<0.001
		Audiovisual quality	−0.212	0.452	<0.001
		Video quality	−0.201	0.457	<0.001
Face factor 7	Outcome measure	Speech quality	−0.270	0.217	0.014
		Effort	0.235	0.522	0.013
		Overall TLX	0.214	0.371	0.013
		Video quality	−0.210	0.122	<0.001
		Audiovisual quality	−0.208	0.125	0.001
		Frustration	0.207	0.496	0.026
		Mental demand	0.200	0.433	0.010
Mean fixation duration	Outcome measure	Audiovisual quality	0.368	0.001	0.022
		Video quality	0.350	0.001	0.022
		Speech quality	0.344	0.001	0.025
Mean pupil dilation	Outcome measure	Mental demand	−0.372	0.006	0.008
Scr delta forwards	Outcome measure	Mental demand	−0.253	0.875	<0.001
		Effort	−0.210	0.146	<0.001

5.2 Quality Evaluations of Actual Conversations

The conversation simulation paradigm has proven to be valuable for testing the IAE model's relationships through the use of short video clips in a passive viewing or listening context. However, the statistical analyses conducted so far serve merely as an intermediate step in validating these relationships. Given the large number of statistical tests performed without alpha level correction, it is reasonable to anticipate a significant proportion of false positives. Consequently, I will apply the same validation approach to a different and independent dataset, transitioning from a passive scenario to an active one.

This section presents the analyses from the actual conversation study paradigm and the results of the experiments. The experimental setup, procedure, and task will be discussed, supplementing the information provided in Sect. 4.1.2.

5.2.1 Experiment and Methods

The dataset utilized for the forthcoming model validation was amassed over three study periods, spanning October to November 2022, February to April 2023, and May to June 2023. All three studies adhered to the same experimental procedures, quality settings, and tasks, yet they differed in terms of the questionnaires employed, technical procedures—such as administering questionnaires on paper or digitally—and the technical stability of the experimental setup. This latter distinction necessitates the use of data exclusively from the last study period for the ensuing analyses, because only the most mature experimental software allowed to record complete trials, without losing data for similar reasons, as described in Sect. 5.1.2. However, to assess straightforward, linear relationships across the model's components, the data from all three study periods were consolidated.

5.2.1.1 Participants

A total of 127 participants were enlisted for the conversational experiments, primarily through the TU Berlin university participant database, supplemented by a handful of participants recruited through an online classifieds website. Of the participants, 67 identified as female and 60 as male, with no other gender specifications provided. The median age was 28 years (IQR: 11y, range: 19–63y). German was indicated as the first language by 79.8% of participants; the remainder were fluent in German. Sixty-eight participants held at least one academic degree, and 50.4% were currently enrolled as students. Participants were awarded a fixed compensation of 30 Euros for their participation, with the experiment averaging 2 hours in duration. The study protocol was approved by the Ethics Committee of Technical University Berlin, Faculty IV Electrical Engineering and Computer

Science (ethics ID: *22_216*), and was conducted in accordance with the Helsinki Declaration. All participants provided informed consent in written form, and their agreement with the procedure was confirmed by signature prior to the start of the recordings.

5.2.2 *Observations*

In total, the dataset obtained from the conversation paradigm consists of 1,140 trials from 127 participants. Similar to the conversation simulation, not every trial has records for all intended variables, resulting in variations in sample sizes. The combination with the least number of samples includes 173 samples from 23 unique participants, while the dataset with the most relationships contains 995 trials from 121 participants.

The observations for the model relationship validation follow the same schema as introduced for the conversation simulation data, with the four perspectives of data flow through the model being discussed and the most significant relationships listed. However, the entire dataset of conducted tests, as well as a list of all significant relationships, is available through the Open Science Framework.[2] In total, 4,869 relationships were tested, with 627 of them proving to be significant, thus representing 12.88% of the total. Among the 183 variables functioning as predictors in the subsequent tests, 121 exhibited significant impacts on other variables within the model's postulated relationships. Conversely, only four variables were found to not be statistically significantly influenced by any of the predictors tested in the following analyses. Importantly, this second set of experiments incorporates various additional variables, as elaborated upon in Chap. 4. These variables include features derived from ECG data, dimensions of conversation interactivity, and the inclusion of several new subjective questions. Consequently, there is a greater number of tests to be conducted compared to the conversation simulation paradigm.

5.2.2.1 Quality Degradation Influences Experience Metrics, States, and Reactions

This initial perspective focuses on the influence of diverse degradation conditions on all terminal components of the model: *dependent states*, *experience ratings*, and an individual's physiological and behavioral *reactions*. For this study, four distinct conditions were employed. The flow through the model is illustrated in Fig. 5.2. Altogether, 79 relationships were examined, with 35 of them found to be significant (44.30%); these are itemized in Table 5.9.

[2] https://osf.io/h3uzr

Table 5.9 Summary of significant correlations between the degradation level of the stimuli and the three classes of variables directly impacted by these stimuli

Predictor	Outcome type	Outcome	β	SE	p
Degrad. lvl	Dependent state	Preference to abandon	−0.546	1.505	<0.001
		Valence	−0.489	0.005	<0.001
		Warmth	−0.252	0.330	<0.001
		Closeness	−0.251	0.047	<0.001
		Competence	−0.139	0.422	<0.001
	Outcome measure	Audiovisual quality	−0.788	0.023	<0.001
		Speech quality	−0.778	0.035	<0.001
		Video quality	−0.769	0.035	<0.001
		Effort	0.683	0.345	<0.001
		TLX frustration	0.383	0.125	<0.001
		TLX effort	0.309	0.138	<0.001
		TLX score	0.265	0.097	<0.001
		TLX performance	0.247	0.138	<0.001
		TLX physical demand	0.198	0.117	<0.001
		TLX mental demand	0.112	0.143	<0.001
	Reaction	Mean pauses speaker change	0.460	0.013	<0.001
		Sojourn mutual silence	0.412	0.011	<0.001
		Prob mutual silence	0.326	0.004	<0.001
		Mean fixation duration	−0.279	2.114	<0.001
		Speaker alternation rate	−0.257	0.119	<0.001
		Median fixation duration	−0.239	1.465	<0.001
		Interruption rate	−0.220	0.039	<0.001
		Prob other talk	−0.151	0.003	<0.001
		Prob self talk	−0.143	0.004	<0.001
		Phasic mean	0.101	0.001	0.019
		Sojourn other	−0.090	0.013	0.007
		Double talk rate other	−0.083	0.116	0.015
		Double talk rate self	−0.082	0.116	0.016
		Sojourn self	−0.077	0.013	0.022
		Prob double talk	−0.074	0.003	<0.001
		Pupil dilation rmssd	0.072	0.001	0.003
		Face factor 3	−0.062	0.046	0.031
		Pupil dilation 5%	−0.027	0.006	0.039
		Pupil dilation 95%	0.020	0.006	0.034
		Mean hr	−0.017	0.090	0.029

Figure 5.7 depicts box-plots of all QoE outcome dimensions, while Table 5.9 enumerates all statistically significant correlations. Regarding the QoE outcome variables, all but one item were demonstrably influenced by the degradation levels of the conditions.

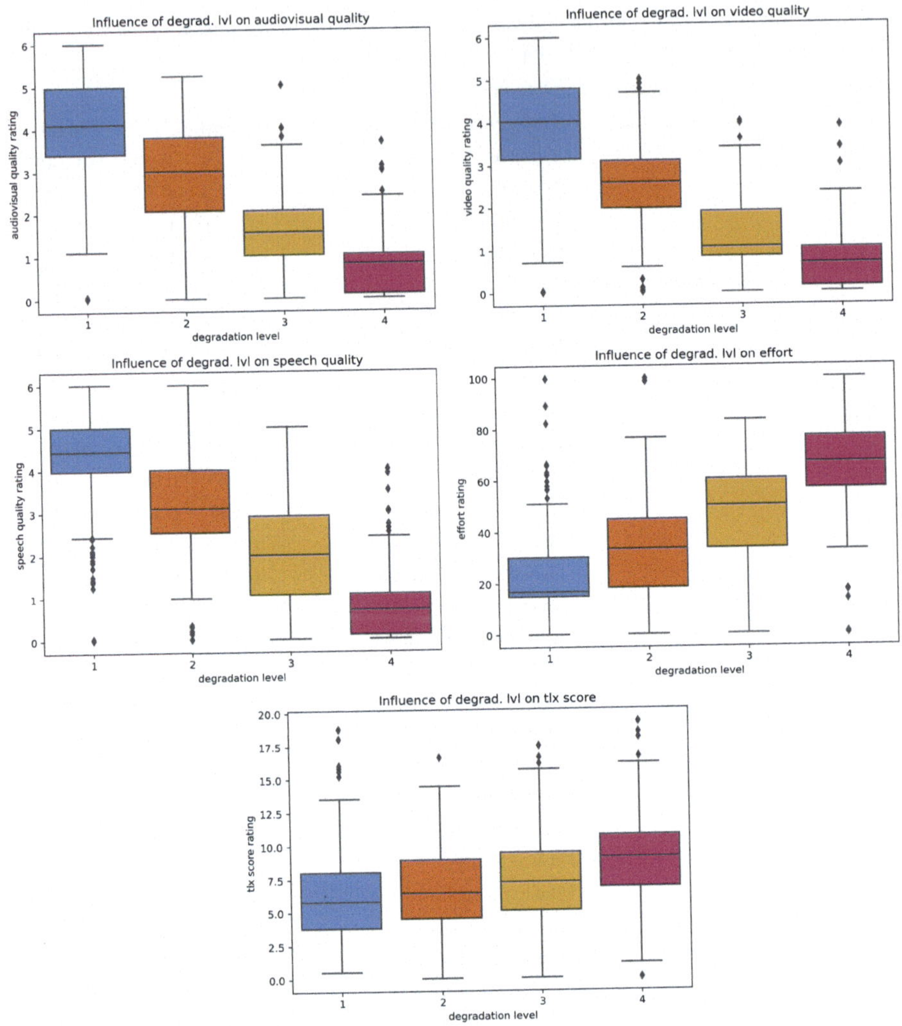

Fig. 5.7 Boxplots of all QoE dimensions influenced by the four degradation conditions in the conversation experiment. From top left to bottom right: the three quality dimensions, followed by the ITU listening/viewing effort item and the overall TLX score

All dependent state variables and all QoE variables, with the exception of arousal and the TLX dimension of temporal demand, were significantly influenced by the degradation level. Notably, the ITU-based listening/viewing effort dimension (detailed in Sect. 4.2) appears to be more significantly impacted by the condition's degradation levels compared to the TLX item that also relates to the perceived effort of the situation. Despite both being termed *effort*, they evidently do not measure the same construct.

The degradation level also influences all categories of reaction measurements. However, we observed that the conversation interactivity dimensions are represented more prominently, while most other dimensions appear infrequently or only once. This could suggest that interactivity dimensions are particularly sensitive to quality changes, in contrast to physiological measures, which seem to respond less precisely to changes across experimental conditions. The widespread impact of the degradation levels on various components of the model and their associated variables signifies a successful manipulation across the conditions implemented in the experiments.

5.2.2.2 Individuality Influences Experience Metrics, States, and Reactions

This subsection scrutinizes the impact of *traits* and *independent states* on *dependent states*, QoE outcome measures, and *reactions*, as depicted in Fig. 5.4. The analyses span a total of 3,239 relationships, 342 (10.56%) of which were deemed statistically significant.

All examined influences were administered only once, while the output variables originating from the subsequent analyses were recorded for each trial. Thus, a significant relationship represents a systematic distinction among participants with specific attributes, independent of any stimuli or quality experience.

Table 5.10 catalogues all statistically significant personality dimensions impacting the terminal measures. It's important to note that the conversation experiment introduced an element that was absent from the simulation experiment: positivity toward technology. All personality dimensions significantly correspond with at least one other variable across the IAE model.

Among frustration-related items, only *emotional intolerance* does not notably influence any terminal variables within the model. Beyond this, all items significantly impact the three model entities *dependent states*, QoE outcome dimensions, and *reactions*, as seen in Table 5.11.

Personal attributes exclusively affect reaction dimensions; no significant correlation with dependent states or QoE outcome measures was identified. Table 5.12 enumerates all significant relationships with an absolute beta value of no less than 0.2.

Lastly, the *independent states* exhibit numerous significant correlations with *dependent states*, QoE outcome dimensions, and the *reaction* dimensions. Table 5.13 provides a summary of these findings.

5.2.2.3 Preceding States, Reactions, and Ratings Influence Subsequent Experience Metrics

Figure 5.5 illustrates the flow of interconnections within the model, focusing on how *dependent states*, *ratings*, and *reactions* influence subsequent QoE outcome measurements. The analysis yielded compelling evidence for carryover effects

Table 5.10 Synopsis of significant correlations between personality dimensions and dependent states, outcome dimensions, and reactions. For improved readability, only relationships with a beta absolute value of at least 0.2 are included

Predictor	Outcome type	Outcome	β	SE	p
Agreeableness	Dependent state	Warmth	−0.224	3.624	0.002
	Reaction	Scr peak rise time	−0.320	0.296	<0.001
Extraversion	Reaction	Speaker alternation rate	−0.274	0.695	<0.001
		Pause rate self	−0.225	0.762	0.003
		Pause rate other	−0.210	0.806	0.010
		Face factor 4	−0.204	0.280	0.004
Neuroticism	Outcome measure	TLX frustration	0.230	0.619	<0.001
	Reaction	Prob double talk	−0.277	0.031	0.004
Openness	Reaction	Prob double talk	0.293	0.034	<0.001
		Mean pauses speaker change	−0.235	0.106	<0.001
		Sojourn mutual silence	−0.235	0.082	0.002
		Double talk rate other	0.234	0.615	<0.001
		Prob mutual silence	−0.221	0.037	0.008
Tech positivity	Outcome measure	Effort	0.352	0.029	<0.001
	Reaction	Pause rate self	0.771	0.004	<0.001
		Pause rate other	0.754	0.004	<0.001
		Prob double talk	−0.679	0.000	<0.001
		Prob mutual silence	0.658	0.000	<0.001
		Speaker alternation rate	0.652	0.005	<0.001
		Sojourn mutual silence	0.532	0.001	<0.001
		Mean pauses speaker change	0.515	0.001	<0.001
		Sojourn double talk	−0.499	0.001	<0.001
		Interruption rate	0.492	0.002	<0.001
		Pupil dilation 5%	0.338	0.001	<0.001
		Double talk rate other	−0.327	0.007	<0.001
		Pupil dilation 95%	0.324	0.002	<0.001
		Double talk rate self	−0.295	0.007	<0.001

between trials. In total, 869 relationships were examined, with 95 exhibiting significance (10.93%). Initially, prior *dependent states* and *reactions* were explored. The significant results are consolidated in Table 5.14, which lists only those relations having an absolute beta value of no less than 0.2. In this context, the cutoff point filters out all dependent state predictors due to their lower beta values. However, prior ratings of closeness, valence, and warmth for or toward the interlocutor, along with the technical assessment of whether to maintain the prior connection or initiate a new call, all demonstrate statistically significant associations with QoE outcome dimensions. The exhaustive list of significant relationships is available in the online repository.

The evaluation of prior QoE ratings on subsequent ratings also indicates potent carryover effects. Notably, task load dimensions frequently appear among the

Table 5.11 Overview of significant correlations between frustration intolerance dimensions and dependent states, outcome dimensions, and reactions. To enhance readability, only relationships with a beta absolute value of at least 0.2 are included

Predictor	Outcome type	Outcome	β	SE	p
Achievement	Dependent state	Warmth	−0.234	2.436	0.002
	Outcome measure	TLX frustration	0.319	0.558	<0.001
		TLX score	0.220	0.424	0.002
Emotional intolerance	Outcome measure	TLX score	0.239	0.412	<0.001
		TLX frustration	0.230	0.534	<0.001
		TLX tempo	0.216	0.513	0.002
		TLX effort	0.213	0.654	0.005
	Reaction	Pause rate other	−0.264	0.610	0.041
		Pause rate self	−0.248	0.486	0.014
Entitlement	Dependent state	Warmth	−0.298	2.333	<0.001
	Outcome measure	TLX frustration	0.257	0.482	<0.001
	Reaction	Sojourn other	−0.215	0.080	0.041
Fairness	Dependent state	Warmth	−0.281	2.093	<0.001
	Outcome measure	TLX frustration	0.232	0.436	<0.001
		TLX score	0.219	0.360	<0.001
	Reaction	Pause rate self	−0.209	0.434	0.041
fds score	Dependent state	Warmth	−0.251	2.660	<0.001
	Outcome measure	TLX frustration	0.280	0.609	<0.001
		TLX score	0.223	0.449	<0.001
		TLX effort	0.206	0.656	0.002

influenced variables, suggesting a greater dependency on prior ratings than on quality ratings, for instance. Table 5.15 compiles all significant associations.

Collectively, both tables corroborate the existence of feedback loops for all three model entities: *dependent states*, *ratings*, and *reactions* are all influenced by an individual's experiences and continue to affect succeeding trials. This supports the bidirectionality of the model's edges and the inclusion of a user's response history in the QoE estimation process.

5.2.2.4 Reactions as a Proxy for Experience Metrics

Finally, in this section, the reactions are evaluated for their predictive potential in deriving the QoE metrics of interest. This perspective is illustrated in Fig. 5.6. A total of 682 relationships were tested, with 155 proving to be significant (22.73%). Table 5.16 exhibits the most potent reaction predictors for the outcome metrics, applying a readability cutoff threshold of $|\beta| = 0.27$. Consistent with previous sections, all statistical results are accessible in the OSF repository for detailed examination. As the table is truncated, only eye-tracking, EDA, and conversation interactivity-related features are presented in the results table, having the strongest

Table 5.12 Summary of significant correlations between personal attributes and dependent states, outcome dimensions, and reactions. Only relationships with a beta absolute value of at least 0.2 are included

Predictor	Outcome type	Outcome	β	SE	p
Age	Reaction	Pupil dilation 95%	−0.544	0.006	<0.001
		Mean pupil dilation	−0.535	0.006	<0.001
		Pupil dilation 5%	−0.490	0.005	<0.001
		pNN50	−0.293	0.171	0.005
		Scr peak count	−0.251	0.059	0.034
		Pupil dilation rmssd	−0.250	0.000	<0.001
		std hr	−0.248	0.032	0.020
		Triangular index	−0.242	0.025	<0.001
		Phasic median	0.240	0.000	<0.001
		lf/hf ratio	0.225	0.028	0.006
		Pause rate self	−0.224	0.025	0.015
Education class	Reaction	Double talk rate self	0.241	0.537	<0.001
Gender identification	Reaction	Min hr	0.215	1.992	0.019
Impaired vision	Reaction	Prob mutual silence	−0.340	0.028	<0.001
		Prob double talk	0.306	0.033	0.002
		Pause rate other	−0.303	0.602	<0.001
		Pause rate self	−0.288	0.539	<0.001
		Mean pauses speaker change	−0.283	0.079	<0.001
		Sojourn mutual silence	−0.270	0.056	<0.001
		Double talk rate self	0.266	0.577	<0.001
		Sojourn double talk	0.224	0.068	0.005
		Mean hr	0.215	2.160	0.003
		Min hr	0.215	2.187	0.009
Occupation class	Reaction	Mean pupil dilation	0.224	0.102	0.011
		pNN50	0.223	2.936	0.018
		Pupil dilation 5%	0.220	0.088	0.010
		Pupil dilation 95%	0.219	0.116	0.013
		Scr peak count	0.208	0.970	0.044
qoe video calls so far	Reaction	rmssd	−0.289	3.856	0.008
		pNN50	−0.255	2.748	0.036
Rom attracted to women	Reaction	Scr peak count	0.251	0.016	0.026
Attraction to interloc sex	Reaction	Prob double talk	0.230	0.027	0.010
		Sojourn double talk	0.215	0.062	0.009
		Pause rate other	−0.201	0.505	0.015
		Prob mutual silence	−0.200	0.025	0.012

(continued)

Table 5.12 (continued)

Predictor	Outcome type	Outcome	β	SE	p
Smoker	Reaction	Scr peak rise time	0.581	0.463	0.002
		Sojourn other	0.264	0.166	0.042
		Double talk rate other	−0.253	0.875	0.006
Using visual aid	Reaction	Sojourn self	0.221	0.073	0.007
Video call dev importance	Reaction	Mean hr	−0.315	0.070	<0.001
Video calls replace f2f	Reaction	Pupil dilation 5%	−0.201	0.002	0.031

relationships of all tests conducted. Nonetheless, all five reaction classes are included in the list of significant predictors: most notably, 16 conversation interactivity variables significantly estimate at least one QoE dimension, followed by eight eye-tracking related features, five facial configuration features, three ECG features, and merely two EDA-based features.

5.3 Meta-Analysis of Both Studies

The analysis of the two studies, namely the simulation-based data analysis and the conversation-based analysis, leads to a convergence of findings. To assess the IAE model's structure, I only consider those relationships that are apparent in both analytical branches, thereby ensuring the validity of the proposed edges and entities of the model, as described in Chap. 3. The design of both experiments was aimed to test all possible connections of the model. Therefore, it follows that all relationships proposed by the model should be discernible in the aforementioned analyses, and hence, both datasets reveal evidence for all connection. Guided by this rationale, only the relationships confirmed by both experiments are considered when accepting or rejecting a proposed relationship.

Nevertheless, the conversation experiments introduced several new variables, which, for example, could not be incorporated in the simulation, such as the conversation interactivity. These dimensions cannot be evaluated in a passive viewing/listening scenario and can only be integrated in a live conversation. For all such instances, a Bonferroni–Holm (BH) alpha level correction was employed on the entire set of conversation-based results. Novel variables, like the interactivity dimensions, are deemed relevant only if they hold significance following the BH correction. Consequently, the resulting set of relationships includes only those that are either confirmed by two independent experiments or abide by the standard procedure of alpha level correction, while still providing statistical evidence. Table 5.17 provides an overview of the number of relationships identified in the analyses of both study paradigms. Furthermore, it lists the number of relationships found for each model entity, validated either through both studies or through the conversation study after alpha level adjustment.

Table 5.13 Summary of significant correlations between independent states and dependent states, outcome dimensions, and reactions. To enhance readability, only relationships with a beta absolute value of at least 0.2 are included

Predictor	Outcome type	Outcome	β	SE	p
Being tired	Reaction	hf power	0.211	5.371	0.035
Cigarette amount	Reaction	Scr peak count	−0.708	0.253	<0.001
		Phasic median	0.411	0.002	0.040
		pNN50	−0.387	0.467	0.001
		Prob mutual silence	−0.358	0.002	<0.001
		Triangular index	−0.350	0.115	0.006
		Scr peak rise time	−0.310	0.041	0.010
		Sojourn other	0.229	0.008	0.006
		hf power	−0.217	13.759	<0.001
Food abstention	Reaction	Face factor 5	0.440	0.124	0.017
		rmssd	0.332	1.684	0.022
		Min hr	−0.276	0.788	0.022
		pNN50	0.276	1.278	0.043
		Mean hr	−0.255	0.736	0.010
		Triangular index	0.237	0.203	0.021
		Prob double talk	−0.232	0.007	0.007
Have had calls with interlocutor	Dependent state	Closeness	0.400	0.501	<0.001
		Competence	0.264	2.408	<0.001
		Warmth	0.254	5.000	0.012
	Outcome measure	TLX performance	−0.209	1.119	0.017
	Reaction	Pupil dilation 95%	−0.237	0.174	0.032
		Mean pupil dilation	−0.236	0.150	0.035
Interlocutor familiarity	Dependent state	Closeness	0.320	0.290	<0.001
		Warmth	0.261	2.598	<0.001
Interlocutor relationship	Dependent state	Closeness	0.363	0.067	<0.001
		Warmth	0.254	0.617	<0.001
		Competence	0.221	0.531	<0.001
	Reaction	Prob double talk	1.119	0.015	<0.001
		Mean pauses speaker change	−0.589	0.027	<0.001
		Sojourn double talk	0.511	0.029	<0.001
Sympathy circle of friends fit	Dependent state	Closeness	0.406	0.006	<0.001
		Warmth	0.262	0.058	0.002
		Competence	0.207	0.059	0.006
	Reaction	Conversational temp	−0.213	0.012	0.025

(continued)

Table 5.13 (continued)

Predictor	Outcome type	Outcome	β	SE	p
Sympathy could be a friend	Dependent state	Closeness	0.435	0.007	<0.001
		Warmth	0.324	0.059	<0.001
		Competence	0.294	0.061	<0.001
	Reaction	Mean hr	−0.735	0.084	<0.001
		Conversational temp	−0.237	0.012	0.013
Sympathy is pleasant	Dependent state	Warmth	0.440	0.070	<0.001
		Competence	0.344	0.096	<0.001
		Closeness	0.340	0.009	<0.001
Sympathy would meet again	Dependent state	Closeness	0.308	0.008	0.004
		Warmth	0.301	0.075	0.002
	Reaction	Conversational temp	−0.214	0.014	0.039

Table 5.14 Effects of prior reactions on QoE outcome measures. Only associations with a beta absolute value of at least 0.2 are included

Predictor	Outcome type	Outcome	β	SE	p
Prev face factor 5	Outcome measure	TLX mental demand	0.233	0.275	0.001
		TLX frustration	0.208	0.236	0.001
Prev face factor 8	Outcome measure	TLX frustration	0.220	0.265	<0.001
Prev interruption rate	Outcome measure	Effort	0.205	1.394	0.001
Prev max hr	Outcome measure	Video quality	−0.202	0.007	0.003
Prev mean hr	Outcome measure	Video quality	−0.220	0.008	<0.001
Prev min hr	Outcome measure	TLX performance	0.210	0.026	0.005
Prev pause rate other	Outcome measure	Effort	0.202	0.835	0.048
Prev pause rate self	Outcome measure	Effort	0.205	0.827	0.044
Prev phasic mean	Outcome measure	TLX tempo	0.690	77.493	<0.001
		TLX mental demand	0.282	6.822	<0.001
Prev pNN50	Outcome measure	TLX performance	−0.250	0.016	<0.001
Prev scr delta backwards	Outcome measure	TLX tempo	−0.207	0.030	<0.001
Prev speaker alternation rate	Outcome measure	Effort	0.238	0.516	<0.001

The *United* column in Table 5.17, indicating the number of relevant relationships per model entity, uncovers two crucial insights: Firstly, every model entity maintains a significant number of relationships with other entities, reflecting the model's structure as hypothesized in Chap. 3. No entity is superfluous as all contribute to the model across various dimensions. Secondly, while most pathways have a small two-digit number of relevant relationships, the *previous trial* relationships distinguish themselves by having very few relationships with other entities. The absolute number of relationships only roughly provides significance as it depends on the number of tests performed. However, it does motivate further investigation to determine whether these connections are essential for the model's predictive performance.

Table 5.15 Impact of previous QoE ratings on following QoE ratings

Predictor	Outcome type	Outcome	β	SE	p
Prev audiovisual quality	Outcome measure	TLX tempo	−0.088	0.102	0.019
Prev video quality	Outcome measure	TLX tempo	−0.090	0.103	0.017
Prev speech quality	Outcome measure	TLX tempo	−0.098	0.095	0.008
Prev effort	Outcome measure	Speech quality	0.123	0.004	0.036
		Video quality	0.119	0.004	0.044
		TLX tempo	0.106	0.007	0.012
Prev TLX effort	Outcome measure	TLX score	0.369	0.064	<0.001
		TLX tempo	0.274	0.045	<0.001
		TLX mental demand	0.252	0.075	<0.001
		TLX frustration	0.167	0.059	0.006
Prev TLX frustration	Outcome measure	TLX frustration	0.321	0.089	<0.001
		TLX tempo	0.202	0.045	< 0.001
		Speech quality	0.145	0.026	0.030
Prev TLX mental demand	Outcome measure	TLX mental demand	0.602	0.043	<0.001
		TLX score	0.486	0.048	<0.001
		TLX effort	0.389	0.076	<0.001
		TLX tempo	0.273	0.056	<0.001
		TLX frustration	0.160	0.064	0.012
Prev TLX performance	Outcome measure	TLX performance	0.216	0.083	0.013
		TLX tempo	0.185	0.049	<0.001
		TLX score	0.126	0.041	0.016
		TLX frustration	0.117	0.057	0.026
		TLX effort	0.114	0.062	0.036
Prev TLX physical demand	Outcome measure	TLX physical demand	0.629	0.036	<0.001
		TLX tempo	0.174	0.052	<0.001
		TLX effort	0.171	0.076	0.007
		Video quality	0.111	0.023	0.040
Prev TLX score	Outcome measure	TLX score	0.545	0.044	<0.001
		TLX effort	0.408	0.119	<0.001
		TLX tempo	0.351	0.074	<0.001
		TLX mental demand	0.317	0.118	<0.001
		TLX frustration	0.232	0.095	<0.001
Prev TLX tempo	Outcome measure	TLX tempo	0.555	0.046	<0.001
		TLX mental demand	0.239	0.077	<0.001
		TLX effort	0.211	0.080	0.002
		TLX score	0.210	0.057	0.003

The findings discussed thus far establish the structure of the model as introduced in Chap. 3. To identify which features among the extensive array of dimensions investigated are pivotal for predicting a person's QoE ratings, the subsequent section will dissect the list of relevant features individually. Although all variables'

Table 5.16 Significant reactions predicting QoE outcome dimensions. Only relationships with a beta absolute value of at least 0.27 are included

Predictor	Outcome type	Outcome	β	SE	p
Mean fixation duration	Outcome measure	Video quality	0.414	0.001	<0.001
		Speech quality	0.411	0.001	<0.001
		Audiovisual quality	0.409	0.001	<0.001
		Effort	−0.396	0.009	<0.001
Median fixation duration	Outcome measure	Audiovisual quality	0.300	0.001	<0.001
		Video quality	0.291	0.001	<0.001
Pupillary activity index	Outcome measure	Effort	−1.684	23.750	0.001
		TLX tempo	1.507	4.422	0.004
		TLX mental demand	1.249	5.044	0.023
		TLX frustration	−0.978	4.522	0.049
Phasic mean	Outcome measure	Effort	0.526	417.999	0.036
Mean pauses speaker change	Outcome measure	Effort	0.650	2.222	<0.001
		Speech quality	−0.536	0.159	< 0.001
		Video quality	−0.463	0.129	< 0.001
		Audiovisual quality	−0.444	0.100	<0.001
		TLX frustration	0.387	0.403	<0.001
		TLX physical demand	0.318	0.442	<0.001
		TLX effort	0.301	0.361	<0.001
Prob double talk	Outcome measure	Effort	−0.308	9.783	<0.001
Prob mutual silence	Outcome measure	Effort	0.574	6.857	<0.001
		Speech quality	−0.490	0.606	<0.001
		Audiovisual quality	−0.448	0.519	<0.001
		Video quality	−0.402	0.507	<0.001
		TLX frustration	0.358	1.620	<0.001
		TLX effort	0.278	1.785	<0.001
		TLX physical demand	0.272	1.246	<0.001
Sojourn mutual silence	Outcome measure	Effort	0.598	2.875	<0.001
		Speech quality	−0.493	0.246	<0.001
		Audiovisual quality	−0.473	0.192	<0.001
		Video quality	−0.424	0.175	<0.001
		TLX frustration	0.395	0.627	<0.001
		TLX effort	0.300	0.648	<0.001
		TLX physical demand	0.291	0.576	<0.001
		TLX score	0.283	0.472	<0.001
Speaker alternation rate	Outcome measure	Video quality	0.373	0.037	<0.001
		TLX frustration	−0.325	0.085	<0.001
		TLX effort	−0.290	0.086	<0.001
		TLX score	−0.271	0.057	<0.001

Table 5.17 The table presents the number of significant relationships per entity of the IAE model. The direction signifies whether the entity acts as a predictor (arrow outwards) or as an outcome (arrow inwards), as three entities are modeled with a bidirectional connection; dependent states, the QoE measures, and the reactions. The "Passive" column enumerates the significant relationships identified in the conversation simulation experiments, while "Active" documents the discovered relationships of the conversation experiments. "United" illustrates the significant relationships either found in both study paradigms or in the conversation experiments after BH alpha level correction

Entity	Direction	Passive	Active	United
Stimulus	Predictor	23 / 49	35 / 79	23
Traits	Predictor	124 / 1,127	256 / 2,291	29
Independent states	Predictor	26 / 196	86 / 948	11
Dependent states	Predictor (prev trial)	15 / 30	6 / 66	1
	Outcome	11 / 84	66 / 252	10
Experience ratings	Predictor (prev trial)	40 / 100	36 / 121	4
	Outcome	150 / 1,130	327 / 2,013	72
Reactions	Predictor (prev trial)	12 / 360	53 / 682	1
	Predictor (direct)	47 / 360	155 / 682	47
	Outcome	126 / 1,008	234 / 2,604	34

connections were important to assess all postulated relationships, the individuality informed QoE estimation process does not involve all model edges simultaneously. The outwards connection from the latent *Experience* entity to the *Dependent States* is not included in the estimation process of one trial. Instead, the *Dependent States* are integrated in the estimation process via their previous values, one trial earlier. This way, they can encode in which state someone was, when they entered a certain quality experience. Likewise, from the QoE ratings, only the previously given responses are included.

5.3.1 *Compilation of Relevant Features*

The compilation of QoE dimensions evaluated in this study, as presented in Sect. 4.2, encompasses quality ratings, task load, and effort ratings. Although the IAE model does not advocate for a specific QoE dimension for evaluation, the carried out tests of the individual relationships, which solely consider one predictor and one outcome, remain particular to the outcome variable. Accordingly, I will organize the predictor variables according to each outcome dimension.

5.3.1.1 Estimation of Quality Experiences

Table 5.18 displays all pertinent variables examined that aid in the estimation of a quality rating. Notably, only one variable, the personality trait of agreeableness,

Table 5.18 Significant variables to estimate a user's quality ratings. Entity names are abbreviated as; S = Stimulus, T = Traits, IS = Independent States, pDS = preceding Dependent States, pER = preceding Experience Ratings, pR = preceding Reactions, and R = Reactions

Entity	Predictor	β	Description
S	Degradation level	−0.895	Quality ratings decrease as the degradation level augments (higher degradation level corresponds to poor quality)
T	Agreeableness	−0.097	As one's agreeability increases, their tendency to rate quality highly diminishes
IS	–	–	–
pDS	–	–	–
pER	TLX frustration	0.145	The higher the level of frustration experienced in a previous trial, the more one is likely to rate the quality higher subsequently
pR	–	–	–
R	Mean fixation duration	0.414	Longer fixation durations correlate with higher quality ratings; suggests enhanced visual focus on the scene during high quality calls
	Interruption rate	0.229	Increased interruptions correlate with higher quality ratings; swift interactions are more probable during high-quality calls
	Mean pauses speaker change	−0.463	Extended pauses between speaker changes are indicative of lower quality; likely due to substantial delays in heavily degraded conditions
	Prob mutual silence	−0.402	A higher likelihood of no one speaking correlates with lower quality ratings
	Sojourn mutual silence	−0.424	Prolonged periods of mutual silence are indicative of lower quality ratings
	Prob self talk	0.225	An increased likelihood of one speaking correlates with higher quality ratings
	Speaker alternation rate	0.373	Similar to the interruption rate, higher alternations signify more interactivity, which is more feasible during high-quality calls

out of the three psychological entities of the IAE model, maintains a statistically significant relationship with quality ratings across both experimental paradigms. Regarding antecedent variables, only the previous frustration value exhibits a statistically significant relationship with quality ratings of subsequent trials. Aside from that, a singular eye-tracking dimension and various conversation interactivity dimensions significantly predict the quality rating of a call.

5.3.1.2 Estimating Effort

Table 5.19 outlines all pertinent variables assessed that facilitate the estimation of an effort rating. Mirroring the quality ratings, only one trait variable, positivity toward technology, demonstrates a statistically significant relationship with effort ratings

Table 5.19 Significant variables to estimate a user's effort ratings. Entity abbreviations are as follows; S = Stimulus, T = Traits, IS = Independent States, pDS = preceding Dependent States, pER = preceding Experience Ratings, pR = preceding Reactions, and R = Reactions

Entity	Predictor	β	Description
S	Degradation level	0.683	As degradation level increases (high degradation level indicates poor quality), the associated effort also rises
T	Tech positivity	0.352	Individuals with a higher positivity toward technology tend to exert more effort, perhaps due to established expectations
IS	–	–	–
pDS	–	–	–
pER	–	–	–
pR	–	–	–
R	Mean fixation duration	−0.308	Longer fixation durations are associated with lesser effort
	Median fixation duration	−0.396	Longer fixation durations are associated with lesser effort
	Prob double talk	−0.266	A higher probability of simultaneous speech correlates with less overall effort
	Mean pauses speaker change	0.650	Longer pauses between speaker changes correlate with a higher overall effort
	Prob mutual silence	0.574	A higher probability of silence from both parties correlates with a higher overall effort
	Sojourn mutual silence	0.598	Extended periods of mutual silence correlate with a higher overall effort
	Prob other talk	−0.254	A higher likelihood that the interlocutor speaks while the user listens correlates with less overall effort

across both experimental paradigms. None of the *Independent States* or preceding variables exhibit influence across both studies. Similar to the quality ratings, the mean fixation duration serves as a predictor for effort, alongside several interactivity variables.

5.3.1.3 Estimating Task Load

Table 5.20 presents all significant variables tested that facilitate the estimation of the overall task load rating. Two trait variables from the FDS instrument concerning one's frustration tolerance show significant relationships with the task load experience. A variable from the preceding experience ratings, the previously perceived temporal demand, a sub-item of the TLX scale, also significantly contributes to the perceived task load in a subsequent trial. Several interactivity variables contribute to the estimation of a task load rating among the reactions.

Table 5.20 Significant variables to estimate a user's task load ratings. Entity abbreviations are as follows; S = Stimulus, T = Traits, IS = Independent States, pDS = preceding Dependent States, pER = preceding Experience Ratings, pR = preceding Reactions, and R = Reactions

Entity	Predictor	β	Description
S	Degradation level	0.481	As the degradation level rises (high degradation level indicates poor quality), the associated task load also increases
T	Achievement	0.220	Higher general task-related frustration (achievement) is associated with an increased task load experience
	Discomfort intolerance	0.190	Higher general discomfort intolerance (demands for easy, comfortable, hassle-free life) correlates with an increased task load experience
IS	–	–	–
pDS	–	–	–
pER	Prev TLX tempo	0.210	Higher temporal demand experienced in a previous trial corresponds with a higher overall task load in the following trial
pR	–	–	–
R	Mean pauses speaker change	0.261	Longer pauses between speaker changes correlate with an elevated task load experience
	Prob mutual silence	0.256	A higher likelihood of mutual silence correlates with an elevated task load experience
	Sojourn mutual silence	0.283	Prolonged periods of mutual silence correlate with an elevated task load experience
	Speaker alternation rate	−0.271	Higher rates of speaker alternation indicate increased interactivity and correlate with a decreased task load experience

5.3.1.4 Significant Influences on Estimator Variables

In total, 17 variables contribute to the estimation of the QoE dimensions addressed in this work, spanning the *Stimulus*, *Traits*, preceding *Experience Ratings*, and the *Reactions* entities. The *Independent States* and *Dependent States* entities do not appear to directly influence any QoE metrics. There is also no evidence from the combined studies for a feedback loop within the *Reactions* entity. However, considering the model's provision for mutual influences across entities, it is worthwhile to investigate if any significant variables previously discussed might be influenced by another variable not yet mentioned. Hence, all pertinent variables exhibiting a significant relationship with one of the 17 QoE estimators are examined.

Table 5.21 enumerates all relevant variables that significantly estimate at least one of the seventeen QoE estimator variables identified earlier. These variables merit consideration as they significantly influence other variables required for estimating individualized QoE metrics. All influenced variables fall under the conversation interactivity metrics of the *Reactions* entity; no other variables were significantly affected by any variables in either of the two studies.

Table 5.21 Relevant variables that influence the seventeen QoE estimator variables. Entity abbreviations are as follows; S = Stimulus, T = Traits, IS = Independent States, pDS = preceding Dependent States, pER = preceding Experience Ratings, pR = preceding Reactions, and R = Reactions; MPSC = mean pauses speaker change, PDT = prob double talk, and PMS = prob mutual silence

Entity	Predictor	Outcome	β	Description
S	Degradation level	MPSC	0.460	As the quality of the stimulus deteriorates, the duration of pauses between speaker changes increases, likely due to increased connection delay
T	Gratification	MPSC	−0.294	Individuals who demand instant gratification tend to have shorter average pauses between speaker changes
	Impaired vision	MPSC	−0.283	Individuals with impaired vision tend to have shorter pauses between speaker changes
	Tech positivity	PDT	0.658	Greater positivity towards technology correlates with a higher probability of double talk
IS	Cigarette amount	PMS	−0.358	The higher the number of cigarettes consumed on the study day, the lower the probability of mutual silence
	Interlocutor relationship	PDT	1.120	The closer the dialogue partners perceive their relationship, the higher the probability of double talk

Table 5.22 Final list of 21 significant variables following the analyses of the two study paradigms

Entity	Variable description	Predictor
Stimulus	Manipulation variable	Degradation level
Traits	Personal attributes	Impaired vision
	Personality	Achievement
		Agreeableness
		Discomfort intolerance
		Gratification
		Tech positivity
Independent states	Biological influences	Cigarette amount
	Social situation	Interlocutor relationship
Experience ratings	Preceding task load rating	Prev TLX frustration
		Prev TLX tempo
Reactions	Eye tracking	Mean fixation duration
		Median fixation duration
	Conversation interactivity	Interruption rate
		Mean pauses speaker change
		Prob double talk
		Prob mutual silence
		Prob other talk
		Prob self talk
		Sojourn mutual silence
		Speaker alternation rate

This analysis introduces four additional variables to the list of relevant ones. Given that two belong to the *Independent State* entity, they should also be included. The final list of model entities, validated using two different study paradigms, therefore includes: the *Stimulus*, *Traits*, *Independent State*, preceding *Experience Ratings*, and the *Reactions*. A comprehensive list of significant variables to consider for individualized QoE estimation, as concluded from this chapter, is provided in Table 5.22.

This chapter exclusively examined simple relationships between two variables, where one variable of an entity was connected by an outgoing arrow to another variable of a separate entity with an incoming arrow. Although this approach was valuable for assessing broad relationships within the model, it overlooked the relationships among variables within the same entity, disregarded colinearity, construct validity of the entity as a whole, and lacks considerations about relationships of variables within entities. Consequently, the analyses conducted thus far serve solely as the initial validation phase.

A subsequent stage must involve an assessment of the entire model, which will be explored in Chap. 6. Such a holistic analysis may result in a different set of variables to consider, given that it would account for inter-variable relationships. The following and the present approaches serve distinct purposes and answer different questions. In order to thoroughly investigate the IAE model, both approaches have been incorporated into this work.

Chapter 6
Model Weights Analysis and Structural Model Refinement

> *Facts are stubborn things; and whatever may be our wishes, our inclinations, or the dictates of our passion, they cannot alter the state of facts and evidence.*
> *John Adams*

This chapter builds upon the validation process initiated in the previous chapter, but adopts a holistic approach by employing Partial Least Squares Structural Equation Modeling (PLS-SEM). The objective is to assess the entire model, rather than focusing on individual relationships. This analysis addresses the fundamental question of whether the proposed model effectively captures the recorded data, thus contributing to the research question of how individuality in the QoE rating process can be modeled.

Additionally, this chapter presents a second key outcome: the determination of weights associated with the proposed relationships. Through subsequent analyses, the strength of these relationships is quantified, alongside the assessment of their effect sizes and statistical significance.

Furthermore, the chapter explores potential modifications to the architecture of the IAE model in order to optimize its alignment with the data, as previously highlighted in the last section of the preceding chapter.

6.1 Data Source

In the previous chapter, the analyses focused on the quality of the stimuli as a predictor variable. However, this assessment was based on the experimental degradation variable, the degradation level, rather than a direct analysis of the stimuli themselves. The degradation level was a discrete score from one (flawless) to four (strongest degradation) and acted as a proxy for the ground truth quality of

R. P. Spang, *Individualized Quality of Experience Estimation in Audiovisual Communication*, T-Labs Series in Telecommunication Services,
https://doi.org/10.1007/978-3-031-67551-5_6

the stimuli, aligning with the experimental conditions during data acquisition. This approach resulted in strong correlations and associations with various other model components (refer to Fig. 5.1). Nevertheless, for practical applications, it is crucial to estimate the quality of the actual stimuli directly. Additionally, considering the focus of this chapter on investigating relationships within and between constructs, it is vital to include direct estimations of stimulus quality instead of the degradation level. Therefore, unlike the previous chapter, the forthcoming analyses will incorporate quality estimations obtained from two instrumental quality predictor models introduced in Sects. 2.1.2.3 and 2.1.2.4.

Furthermore, the subsequent analyses will exclusively utilize data from the conversation experiments. By restricting the analysis to this specific dataset, all relevant variables can be included, such as interactivity, which is not available in the simulation study.

6.2 Structural Equation Modeling

A SEM is a statistical technique used to analyze complex relationships involving multiple variables and constructs. It facilitates understanding of the overall relationship structure and the individual contributions of each variable to the final estimation and allows for simultaneous analysis of multiple variables, considering them as interconnected elements within a system. Hence, it is an appropriate tool for assessing the validity of the proposed IAE model.

Additionally, SEM enables the assessment of measurement models, ensuring that observed variables reliably represent the intended constructs. This is a confirmatory factor analysis, offering a second perspective on variable selection. While Chap. 5 examined each individual potential relationship, the validity assessment of the measurement model within the SEM context examines factor loadings onto separate constructs (referred to as entities in the IAE context), thereby identifying unsuitable variables.

Statistical indices provided by SEM allow for the evaluation of the model's fit to the observed data. These indices assess the extent to which the proposed model aligns with the actual relationships among and between constructs. Moreover, model comparison enables the testing of different structural models against each other. It serves as a valuable tool to determine whether the IAE model's structure truly fits the data better than a baseline model, such as one consisting solely of an instrumental quality estimation model. Furthermore, considering the limited evidence for the hypothesized *Dependent States* construct in the previous chapter, an SEM-based model comparison can determine whether the original model or a reduced version fits the data better.

6.2.1 Basic Architecture

A SEM consists of two main elements: the outer model, also known as the measurement model, and the inner model, referred to as the structural model. The structural model defines the relationships between constructs and represents the discussed IAE model, as it specifies the directed relationships between constructs.

However, one exception should be noted: The IAE model suggests a latent "Experience" entity at its center, which captures the cognitive processes that contribute to an individual's experience. This latent experience serves as the foundation for deriving evaluations and ratings. While this concept is valuable for acknowledging the various and partially unknown cognitive processes, in the context of applied modeling, this entity would be omitted. Including this latent construct in the structural model within the SEM context would not provide additional estimation value. Instead, it would mirror the "Experience Rating" entity and its variables of interest. Therefore, in the structural model within the SEM context, all three psychological constructs and the *Reactions* are directly connected to the *Experience Rating* entity.

The measurement model includes the relations to the data recorded during the experiments and defines which variables belong to which construct. Both together, they describe the variables (often called indicators in the SEM context), indicate the construct to which each variable belongs, and define the relationships between constructs. Figure 6.1 illustrates an example of such a path model, depicting the complete configuration of the IAE model.

To begin with an informed set of variables, all variables for which at least one significant relationship was found in the previous chapter were included in the initial version of the SEM.

6.2.2 Iterative Model Validation

The validity of the model and the refinement of the variable selection are appraised through various tests and indices mentioned in the literature, all of which elucidate the statistical characteristics of the model. These tools facilitate the identification of potential weak points and provide means for subsequent model iterations. The ensuing subsections describe the iterative validation process leading to the final configuration of the SEM.

6.2.2.1 Indicator Reliability

As indicated by Hair Jr et al. (2021), the initial step in assessing a measurement model is to inspect the extent of variance in each construct explained by its respective items, signifying the indicator's reliability. An indicator refers to an

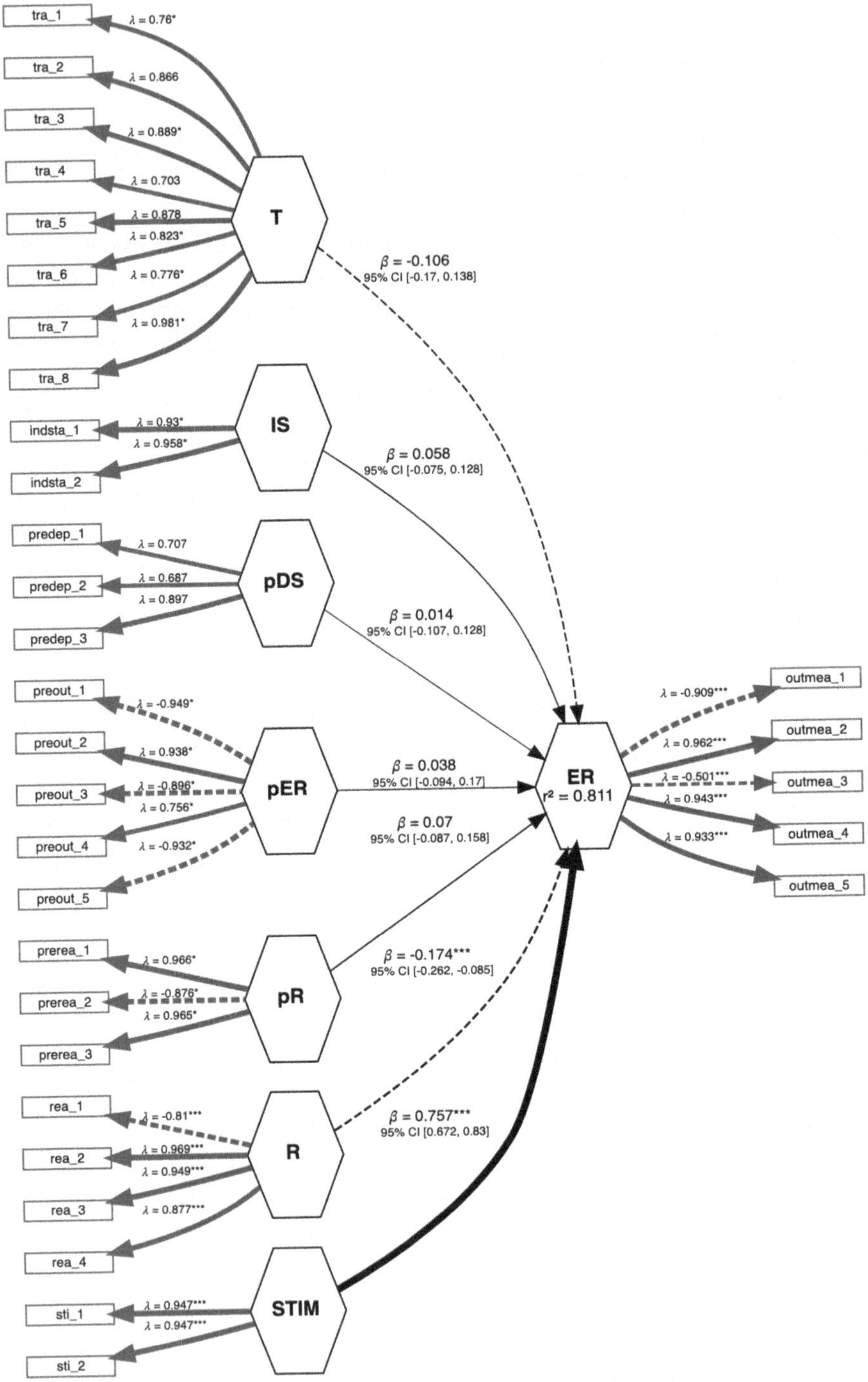

Fig. 6.1 SEM/path model of the full IAE model configuration. Abbreviations of constructs: T = Traits, IS = Independent States, pDS = preceding Dependent States, pER = preceding Experience Ratings, pR = preceding Reactions, R = Reactions, STIM = Stimulus. Abbreviations of variables follow the names of the constructs to which they belong and are explained in detail in Table 6.2

individual variable, such as the variable *neuroticism* from the big five personality traits, belonging to the *Traits* construct. The explained variance of an indicator is obtained by squaring its loading, where loading refers to the bivariate correlation between the indicator and the construct. Therefore, indicator reliability represents an indicator's commonality. Indicator loadings exceeding 0.708 are recommended by Hair Jr et al. (2021), as these suggest that the construct elucidates more than 50 percent of the indicator's variance, providing an acceptable level of indicator reliability. This threshold is handy in selecting an apt subset of variables for each construct. Nevertheless, as contended by Hulland (1999), weaker indicator loadings are frequently utilized, particularly in newly developed instrument contexts. Hair et al. (2022) suggest utilizing 0.4 as a hard cutoff, and scrutinizing all indicators falling within the range of 0.4 and 0.708 to discern whether the internal consistency reliability or convergent validity genuinely amplifies (discussed in the forthcoming section). The authors argue that an indicator should only be discarded if these measures indeed improve; if no improvement is noted, the variable should be retained.

This process was executed iteratively: Commencing with the initial set of variables, the loadings were evaluated and any variable with a loading under 0.4 was omitted. This pruning reduced the total number of indicators to 54. Subsequently, a new model was established using only the retained indicators. Once more, the indicator loadings of this new model were evaluated and checked for any variables exhibiting a loading below 0.4 (since a different set of indicators may cause factors to drop below the cutoff again). In the absence of such cases, the weakest indicator was omitted, a new model was constructed, and its consistency reliability and convergent validity were appraised. If these measures improved, the indicator was permanently excluded; if not, the indicator was reintegrated. This process was repeated for every indicator having a loading between 0.4 and 0.708. The process terminated with a final set of 32 indicators.

Among these 32 indicators, five were outcome variables (audiovisual quality, speech quality, video quality, effort, and task load), along with two stimulus variables encapsulating signal analyses of the video-telephony signal, i.e. algorithmically derived speech and video quality scores describing the stimulus. Taking these into account, only 25 variables were left to describe each participant regarding their traits, states, or reactions.

6.2.2.2 Internal Consistency Reliability

An evaluation of the internal consistency reliability, which measures the degree of association among indicators of the same construct, was conducted. As per the recommendation of Jöreskog (1971), the composite reliability, denoted as rho_c, should exceed a threshold of 0.6, with higher values indicating superior reliability. Another widely utilized metric is the consistent reliability coefficient, denoted as rho_A (Dijkstra, 2009; Dijkstra & Henseler, 2015).

Table 6.1 Comparison of internal consistency reliability by construct

Construct	rho_c	rho_A	AVE
Stimulus (STIM)	0.946	0.886	0.897
Traits (T)	0.949	0.956	0.703
Independent States (IS)	0.943	0.917	0.891
Preceding Dependent States (pDS)	0.811	0.955	0.592
Preceding Experience Ratings (pER)	0.545	0.962	0.804
Preceding Reactions (pR)	0.752	0.970	0.878
Reactions (R)	0.843	0.927	0.817
Experience Ratings (ER)	0.623	0.952	0.753

The reliability analysis conducted on the 32 variables identified in the initial step provided satisfactory reliability values, except for the feedback loop of previous experience ratings (refer to Table 6.1). While the rho_c of the *Preceding Experience Ratings* is slightly below the threshold of 0.6, its rho_A value implies reliable consistency. Thus, it is deemed acceptable to retain this construct in the model.

6.2.2.3 Convergent Validity

Convergent validity, which measures the extent to which a construct converges in explaining the variance of its indicators, was subsequently assessed. The Average Variance Extracted (AVE) is utilized as the metric for this evaluation. The AVE represents the mean of the squared loadings of the indicators associated with a construct, essentially being the communality of the construct. The AVE is considered acceptable if it is above 0.5, implying that the construct explains 50 percent or more of the variance of its indicators (Hair et al., 2022). As shown in Table 6.1, the AVE scores for all constructs surpass this threshold.

6.2.2.4 Evaluation of Discriminant Validity

Discriminant validity assesses the extent to which a construct is uniquely distinguishable from other constructs in the structural model. The Heterotrait–Monotrait (HTMT) ratio of correlations is a recommended tool for this analysis, as proposed by Hair Jr et al. (2021) and Henseler et al. (2015). This ratio represents the mean value of the correlations of indicators across constructs, relative to the geometric mean of the average correlations of the indicators measuring the same construct.

The existence of discriminant validity issues is indicated by high HTMT values. A threshold value of 0.85 has been proposed by Henseler et al. (2015). In the current model, no construct pair exceeds this criterion, implying satisfactory discriminant validity.

6.2.2.5 Reliability and Validity Checks: Conclusion

Employing the evaluation strategy put forth by Hair Jr et al. (2021), the original list of variables was refined to 32 (inclusive of 25 values related to individuality) to assure indicator reliability. With this refined model, all reliability and validity indices demonstrate their appropriateness for the data in question.

The identified variables are enumerated in Table 6.2. A visualization of the complete model is provided in Fig. 6.1.

Table 6.2 List of the 32 variables post iterative indicator validation for the complete IAE model

Entity	Variable	Loading	Abbreviation
Stimulus (STIM)	Mapped NISQA MOS	0.947	sti_1
	Mapped DOVER MOS	0.947	sti_2
Traits (T)	Fairness	0.760	tra_1
	Achievement	0.866	tra_2
	Discomfort intolerance	0.889	tra_3
	Neuroticism	0.703	tra_4
	Emotional intolerance	0.878	tra_5
	Gratification	0.823	tra_6
	Entitlement	0.776	tra_7
	FDS score	0.981	tra_8
Independent State (IS)	Interlocutor familiarity	0.930	indsta_1
	Sympathy circle of friends fit	0.958	indsta_2
Preceding Dependent State (pDS)	Prev closeness	0.707	predep_1
	Prev warmth	0.687	predep_2
	Prev valence	0.897	predep_3
Experience Rating (ER)	Effort	−0.909	outmea_1
	Audiovisual quality	0.962	outmea_2
	TLX score	−0.501	outmea_3
	Speech quality	0.943	outmea_4
	Video quality	0.933	outmea_5
Preceding Experience Rating (pER)	Prev speech quality	−0.949	preout_1
	Prev effort	0.938	preout_2
	Prev video quality	−0.896	preout_3
	Prev TLX frustration	0.756	preout_4
	Prev audiovisual quality	−0.932	preout_5
Preceding Reaction (pR)	Prev mutual silence sojourn	0.966	prerea_1
	Prev speaker alternation rate	−0.876	prerea_2
	Prev mean pauses speaker change	0.965	prerea_3
Reaction (R)	Speaker alternation rate	−0.810	rea_1
	Sojourn mutual silence	0.969	rea_2
	Mean pauses speaker change	0.949	rea_3
	Prob mutual silence	0.877	rea_4

6.2.3 Comparison of Models

Leveraging the model that has been designed and validated above as a basis, we can undertake a comparison of models to discern the optimal one for the recorded data. There are three candidates to consider:

- The baseline model is composed only of scores from the instrumental quality estimation model. This embodies the traditional approach where cutting-edge quality estimation models are utilized to score the speech and video signal. Collectively, these scores form the construct of the *Stimulus*, which is subsequently used to estimate the defined QoE dimensions.
- The complete IAE model as presented in Chap. 3. This model includes all entities discussed in preceding chapters and corresponds to the one modeled earlier (see Fig. 6.1).
- A pared-down version of the IAE model. The need for this option emerged in the previous chapter due to a lack of evidence for a *Dependent States* entity in the combined analysis of both experimental paradigms.

As per Sharma et al. (2021), the Bayesian Information Criterion (BIC) facilitates a balanced trade-off between model fit and predictive capability in the estimation of Partial Least Squares (PLS) path models. The model yielding the lowest BIC value is preferred. However, this criterion fails to provide insight into the relative evidence weights in favor of the models being considered. To address this, Akaike weights are used to gauge a model's relative likelihood of being "the data generation model," given the data and a set of rival models (Danks et al., 2020). Higher Akaike weights suggest that the chosen model likely better represents the data generation model.

Initially, all three models are created using the same dataset, only the structural model varies. Starting from the full model discussed earlier, the reduced model simply excludes the *Dependent States* entity, and the baseline model retains only the *Stimulus* and the *Experience Rating* entity. Utilizing both criteria, the BIC and the Akaike weights, I select the superior model from the three candidates.

The baseline model boasts a BIC value of −292.491 and an Akaike value of 0.021. Conversely, the full model, as hypothesized in Chap. 3, has a BIC of −291.879, and its Akaike value stands at 0.015. The reduced model, as suggested by the data analysis in Chap. 5, possesses a BIC value of −300.096 along with an Akaike weight of 0.962. This analysis evidently favors the reduced model over the other two options. Thus, the structure of the IAE model is settled; the initially hypothesized structure is rejected as the statistical analyses advocate a simplified version of the model for better data fit. The alterations and the ultimate structure of the model are consolidated in Sect. 6.3. This section also incorporates an analysis of the final weights of the constructs, demonstrating the power of the relationships each construct maintains with *Experience Ratings*.

6.3 The Final IAE Model

The model comparison yielded an updated structure for the IAE model that best explains the data variability. This aligns with the final outcomes of the relationship analysis in Chap. 5. The new structure discards the *Dependent States* entity and the feedback loop relationship from the *Reactions* to the latent *Experience* entity. Figure 6.2 illustrates this revised structure, and Appendix C, Fig. C.1 displays the updated model in the estimation application. This chapter concludes with an exploration of the statistical properties of this revised model in the context of an SEM analysis.

6.3.1 *Variables and Weights*

The iterative model validation process delineated in Sect. 6.2.2 was replicated to examine the edge weights of the revised, pared-down model. All validation steps were reiterated as described earlier, ensuring that all validity and reliability thresholds were met. Figure 6.3 depicts the revised path model.

Bootstrapping was employed to compute confidence intervals around the estimated parameters. In accordance with the recommendations of Hair Jr et al. (2021), 10,000 resamples were generated. Bootstrapping resamples from the original data to account for sample variability and creating intervals that reflect the uncertainty in the estimates. Confidence intervals offer a range of plausible values for the parameters,

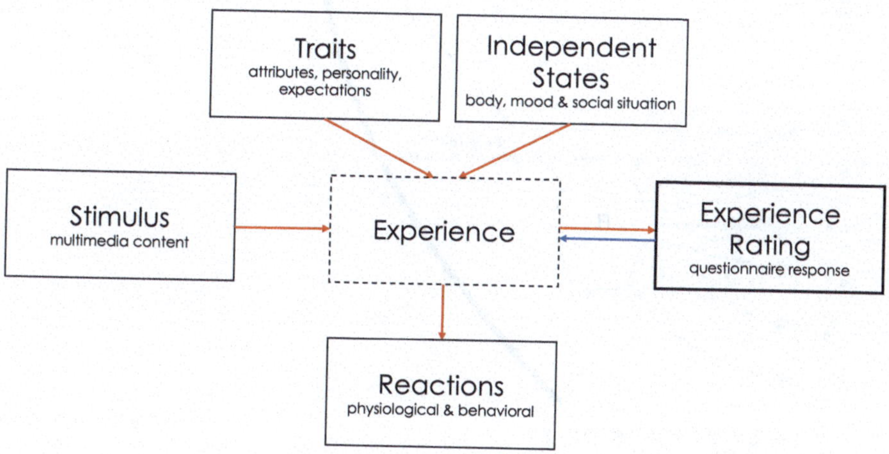

Fig. 6.2 Revised model structure featuring only entities and relationships substantiated by experimental evidence

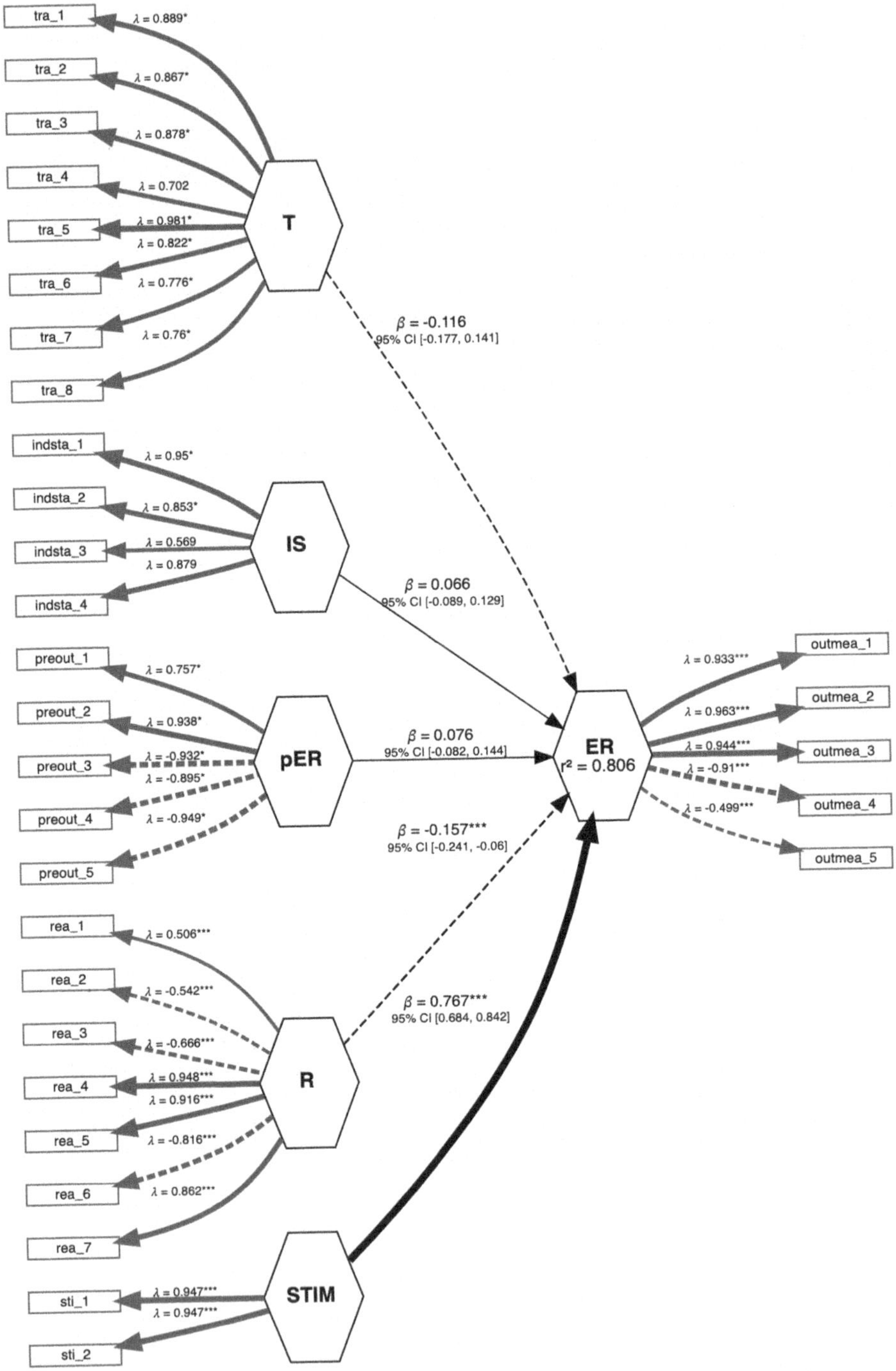

Fig. 6.3 SEM/path model of the streamlined IAE model configuration. Abbreviations of constructs: T = Traits, IS = Independent States, pER = Preceding Experience Ratings, R = Reactions, STIM = Stimulus. The abbreviations of variables correspond to the names of the construct they belong to and are elaborated in detail in Table 6.2. Line thickness indicates the strength of the relationship; dashed lines indicate negative relationships

providing insight into the precision and reliability of the estimates, and enabling the derivation of statistical significance through t-values (significance is reached if $|t|$ exceeds 1.96).

The resultant model yields an $adj.R^2 = 0.801$. The β values, which indicate the strength of the relationship between constructs, are highest for the *Stimulus* construct with $\beta = 0.767$. The $f2$ effect size of this predictor construct on the *Experience Ratings* is notably large at $f2 = 1.601$. General guidelines suggest 0.02, 0.15, and 0.35 for small, medium, and large effect sizes (Cohen, 2013). The bootstrapped model assigns a $t = 19.055$ to this relationship, indicating a significant association between the *Stimulus* and the *Experience Ratings*. The *Reactions* have a $\beta = -0.157$ loading, a small effect of $f2 = 0.060$, which is still statistically significant at $t = -3.400$. The *Traits* have the next strongest relationship with the outcome ratings: with a loading of $\beta = -0.116$, a small effect size of $f2 = 0.057$, and a t-value $\in [-1.96, 1.96]$, ($t = -1.401$), this relationship is not statistically significant. The previous *Experience Ratings* show a beta loading of $\beta = 0.076$, a small effect of $f2 = 0.029$ and are not statistically significant ($t = 1.530$). Lastly, the *Independent States* have the lowest beta loading of $\beta = 0.066$, a small effect of $f2 = 0.022$, and are not statistically significant ($t = 1.217$).

The final variable set identified for the streamlined SEM is listed in Table 6.3.

6.4 Discussion

The initial hypothesis suggested that *Dependent States* would contribute to experience; however, the recorded data did not support this. Analysis revealed a better data/model fit when omitting this entity. Similarly, the evidence did not support the feedback loop of *Reactions*. Consequently, the model included only one feedback loop that allowed experience ratings of a preceding trial to influence the subsequent one. This mechanic accommodated carryover effects.

The final model, refined in Chap. 6, distinguished between *Traits* and *States* psychologically, recognized *Reactions* as an outlet of internal *Experiences*, and included one feedback loop for *Experience Ratings*. The final structure is shown in Fig. 6.3.

After conducting relationship analyses and employing structural equation modeling (SEM), I found that only conversational interactivity dimensions remained within the *Reactions* construct. Other physiological and behavioral variables either displayed inconsistent influences on other factors or were eliminated during the indicator reliability test (Sect. 6.2.2.1). However, the relationship analyses confirmed that the *Reactions* construct primarily contained interactivity variables. This duality provides a meaningful advantage as it allows a direct discussion about the decision to group all *Reactions* variables together.

Table 6.3 List of the 26 variables after the iterative indicator validation for the streamlined SEM

Entity	Variable	Loading	Abbreviation
Stimulus	Video quality score	0.947	sti_1
	Speech quality score	0.947	sti_2
Traits	Discomfort intolerance	0.889	tra_1
	Achievement	0.867	tra_2
	Emotional intolerance	0.878	tra_3
	Neuroticism	0.702	tra_4
	FDS score	0.981	tra_5
	Gratification	0.822	tra_6
	Entitlement	0.776	tra_7
	Fairness	0.760	tra_8
Independent State	Interlocutor familiarity	0.950	indsta_1
	Have had calls with interlocutor	0.853	indsta_2
	Sympathy is pleasant	0.569	indsta_3
	Sympathy circle of friends	0.879	indsta_4
Experience Rating	Video quality	0.933	outmea_1
	Audiovisual quality	0.963	outmea_2
	Speech quality	0.944	outmea_3
	Effort	−0.910	outmea_4
	TLX score	−0.499	outmea_5
Preceding Experience Rating	Prev TLX frustration	0.779	preout_1
	Prev effort	0.928	preout_2
	Prev audiovisual quality	−0.919	preout_3
	Prev video quality	−0.878	preout_4
	Prev speech quality	0.557	preout_5
Reaction	Sojourn double talk	0.506	rea_1
	Prob other talk	−0.542	rea_2
	Interruption rate	−0.666	rea_3
	Sojourn mutual silence	0.948	rea_4
	Mean pauses speaker change	0.916	rea_5
	Speaker alternation rate	−0.816	rea_6
	Prob mutual silence	0.862	rea_7

Despite expecting a broader range of physiological variables in the final selection, the focus on interactivity variables was advantageous: they quantified user behavior while being relatively easy to record and compute. Their accessibility and strong predictive value make them valuable in this context.

Chapter 7
Comparative Performance Analysis: Advantage of IAE Over Signal-Only Models

Extraordinary claims require extraordinary evidence.
Carl Sagan

The effectiveness of the IAE model, and its accuracy in QoE estimation compared to conventional methods that do not include individuality information, is examined in this chapter. The methodology involves comparing IAE's performance against two state-of-the-art instrumental quality estimation models, which score the speech and video quality of recorded video-telephony conversations. The difference between users' QoE ratings and the algorithmically derived scores provides a measure of estimation accuracy, which an effective model should aim to minimize. To ascertain whether incorporating individuality information, as proposed by the IAE model, enhances estimation accuracy, an Individuality Assisted Estimation process is also conducted and evaluated. If the IAE model yields smaller errors than the instrumental-only approach, it can be considered more favorable and effective in accounting for additional variance by incorporating user knowledge.

The results from the model comparison of the SEM in Sect. 6.2.3 have already suggested that the individuality-informed model is superior to the instrumental-only approach. The challenge, however, is quantifying the extent of this improvement and determining its practical significance.

7.1 Experiments

This section delves into the experiments' design and the preliminary considerations that were taken into account. These include scale conversions, the signal-based quality analysis of speech and video dimensions, the derivation of a joint audiovisual quality estimation, the use of instrumentally estimated quality scores for other QoE metrics, the dataset, and the statistical methods used to investigate estimation accuracy.

R. P. Spang, *Individualized Quality of Experience Estimation in Audiovisual Communication*, T-Labs Series in Telecommunication Services,
https://doi.org/10.1007/978-3-031-67551-5_7

7.1.1 Signal-Based Quality Analyses

The IAE model, rather than suggesting an analysis of the stimuli itself, describes how to incorporate additional personal information into a signal-based quality estimation. This allows the estimator system to base its estimations on both the signal-based analysis and the individuality information. To assess the added value of incorporating individuality data, the same instrumental quality estimation models are utilized in both the baseline and the IAE model configurations.

As detailed in Sects. 2.1.2.3 and 2.1.2.4, NISQA was employed as a state-of-the-art speech quality estimation model, while DOVER was used as a top-tier video quality estimation model.

Both quality estimators generated results on different scales than those used by users to rate quality after each trial. Specifically, the instrumental speech quality model returned quality scores in the conventional one to five range, whereas the instrumental video quality model produced scores between 0 and 100. However, in the conducted experiments, a seven-point scale (five points plus overflow), as described in Sect. 4.2, was employed. To enable comparison between these estimations and actual ratings, a mapping process was executed.

While it is common practice to map estimated scores to users' ratings to eliminate subjective test biases during objective model performance comparisons and across numerous subjective tests (ITU-T Recommendation P.1401, 2020), these usually map estimations to the average scores obtained from multiple users. In conventional listening/viewing tests, many participants rate the same stimulus, and therefore for one stimulus, many scores are recorded. This process reduces variability by eliminating variance introduced by individuals through averaging. However, in this study, one stimulus (one side of one conversation) was only rated by a single individual. Therefore, the score mapping employed here aimed to better align the estimation to match all the users' ratings.

In alignment with Sheikh et al. (2006) and others, such as Zhang et al. (2015) and Ying et al. (2023), who used the same methodology in similar cases, the parameters of a nonlinear function were estimated. This function is designed to recalibrate the predicted values, aligning them with the scale used for subjective scores. The nonlinear regression in Eq. 7.1 deploys a mapping function with β_i parameters (where $i = 1, 2, \ldots, 5$) that need to be fitted.

$$\text{Quality}(x) = \beta_1 \left(0.5 - \frac{1}{1 + e^{\beta_2 (x - \beta_3)}}\right) + \beta_4 x + \beta_5 \tag{7.1}$$

After fitting the parameters and applying the resultant mapping function to the outputs from both instruments, the values from both systems were effectively matched to the same range as the user ratings.

7.1.2 *Model Fine-Tuning*

To optimize the performance of the instrumental quality estimation models on the custom datasets created for this thesis, both models underwent a fine-tuning process. Fine-tuning involves adjusting pre-trained models to enhance their performance on a specific task or dataset. Initially, the preexisting models were trained on a large general dataset. During fine-tuning, the weights of the pre-trained models were partially frozen to retain the knowledge acquired from the general dataset. Subsequently, the models were trained on the new dataset using a smaller learning rate. This allowed the models to adapt and specialize their learned features to the specific characteristics of the new data, resulting in improved performance on the target task.

Section 5.2.1 explained that many of the recorded trials had missing values due to incomplete questionnaires, sensor or software failures, or participants obstructing the face-tracking system. However, all incomplete trials that still had labels for their speech or video quality experiences were used for the fine-tuning process. Approximately two-thirds of the total recorded data were incomplete but proved valuable for the retraining process of the models. This process allowed the weights of the models to be updated and used to score the remaining complete trials. The large portion of incomplete data initially seemed problematic, but it was successfully utilized to ensure comparable performance with state-of-the-art quality estimation models in their respective fields.

Even though the models were fine-tuned explicitly for this specific task, a third-order mapping between the signal-based quality estimation and the user's ratings, as discussed in ITU-T Recommendation P.1401 (2020) (Scale calibration of objective quality models), was performed in addition to the rescale mapping discussed previously. This was to ensure fairness when comparing the signal-only approach with the IAE fusion model, which would inherently assign weights and, depending on its architecture and functioning, a random intercept to the scores to better match the outcome.

7.1.2.1 Speech Quality Estimation

The NISQA model, a comprehensive speech quality evaluator, was utilized and fine-tuned using the conversation experiment dataset referenced in Sect. 5.2.1. The source code and the pre-trained model version 2.0 were utilized for this purpose. The entire model was updated during the fine-tuning phase, indicating a thorough retraining process involving all layers of the model (Mittag et al., 2021).

For fine-tuning, 425 speech recordings from the conversation experiments, each 120 seconds long, were utilized, with each speech quality rating given by the participants serving as a target score. The retraining process spanned 19 epochs with a batch size of four samples, a training to validation data ratio of 80:20, and utilized the ADAM optimizer with a learning rate of 0.001. Early stopping was permitted if no improvement was observed in validation RMSE or correlation r_p for 20 consecutive epochs.

7.1.2.2 Video Quality Estimation

The DOVER model was utilized as a signal-based video quality estimation model, and it was fine-tuned using the same portion of the dataset from the conversation experiments aforementioned. The source code and the pre-trained model version 0.5.0 were employed in this process.

To fine-tune the model, new regression heads were trained using 425 video clips from the experiments. Each of these clips was technically identical to the test clips used later in this chapter's analysis, but they displayed one or more incompleteness in the subjective or objective recordings captured during the same trial.

The DOVER model is unique in that it produces a *technical* quality score and an *aesthetics* quality score, which it averages to generate a "final score." This final score was used as the video quality metric in the subsequent sections of this chapter.

7.1.3 Data Source and Metrics of the Performance Estimation

The data used for the performance evaluation are derived from the conversation experiment dataset, as described in Chap. 6. This dataset comprises 190 trials from 28 participants, with complete information on variables, sensors, and estimations.

To assess the models' performance, several metrics are employed, following the suggestions by Ying et al. (2023). These metrics include Spearman's Rank Order Correlation Coefficient (SROCC), Pearson's Linear Correlation Coefficient (PLCC), and the Root Mean Square Error (RMSE) to measure the models' performance. Additionally, also Mean Average Error (MAE), $adj.R^2$, and R^2 scores are computed, as these are commonly used in the literature and provide straightforward interpretation.

SROCC measures the rank consistency between the predicted and subjective scores, and is given by:

$$\rho = 1 - \frac{6 \sum d_i^2}{n(n^2 - 1)} \tag{7.2}$$

Here, ρ represents the SROCC, d_i refers to the rank difference between corresponding pairs of observations, and n represents the total number of observations. It is important to note that this formula assumes that the observations have already been ranked.

PLCC quantifies the linear correlation between the predicted score and the subjective score, expressed as:

$$r = \frac{\sum (x_i - \bar{x})(y_i - \bar{y})}{\sqrt{\sum (x_i - \bar{x})^2 \sum (y_i - \bar{y})^2}} \tag{7.3}$$

In this equation, r denotes the PLCC, x_i and y_i denote the quality scores (objective and subjective), and $\bar{x}$ and $\bar{y}$ represent the means of the observations in the two variables. The summations are taken over all observations in the dataset. The numerator calculates the covariance between the variables, while the denominator computes the product of their standard deviations, resulting in Pearson's correlation coefficient.

RMSE measures the average squared difference between predicted values and actual values, serving as a widely used metric for evaluating regression models. It is calculated as:

$$\text{RMSE} = \sqrt{\frac{1}{n}\sum_{i=1}^{n}(y_i - \hat{y}_i)^2} \tag{7.4}$$

Here, n denotes the total number of observations, y_i represents the subjective quality rating for the i-th data point, and $\hat{y}_i$ represents the estimated quality for the i-th data point. The summation is taken over all n data points.

MAE measures the average absolute difference between the estimates and the rated values, indicating the closeness of the estimations to the actual values:

$$\text{MAE} = \frac{1}{n}\sum_{i=1}^{n}|y_i - \hat{y}_i| \tag{7.5}$$

In this formula, y_i represents the user's rating for the i-th observation, and $\hat{y}_i$ represents the estimated value for the i-th observation. n represents the total number of trials. Unlike RMSE, MAE provides a balanced view of the overall error magnitude.

Lastly, R^2 (coefficient of determination) indicates the proportion of the variance in the dependent variable explained by the independent variables in a regression model. The adjusted R^2 accounts for the number of predictors and sample size, penalizing models with many predictors. The formulas for adjusted R^2 and R^2 are as follows:

$$R^2 = 1 - \frac{\sum_{i=1}^{n}(y_i - \hat{y}_i)^2}{\sum_{i=1}^{n}(y_i - \bar{y})^2} \tag{7.6}$$

$$adj.\ R^2 = 1 - \left(\frac{n-1}{n-k-1}\right)\left(1 - R^2\right) \tag{7.7}$$

Here, n represents the total number of trials, k is the number of predictors (independent variables) in the model, y_i denotes the actual value of the estimated quality for the i-th observation, $\hat{y}_i$ represents the subjective quality rating for the i-th observation, and $\bar{y}$ represents the mean of the quality estimations.

7.1.4 Application of the IAE Model

To generate estimations that incorporate individuality information of users, a leave-one-subject-out cross-validation approach is followed. A regressor is trained as a residual learner, aiming to estimate the difference between a user's actual rating and the model's estimation.

For example, when estimating the subjective speech quality for user A, the difference between the NISQA score and the actual ratings of all other users is computed. The regressor is then trained to learn this difference using all available data except for the trials involving user A. Subsequently, the regressor is applied to user A's data to estimate the error that an instrumental quality estimation model would make. This estimated error serves as an offset and is added to the NISQA scores to obtain the final estimation for user A. The rationale behind residual learning is that while objective models aim to estimate an MOS averaged over a larger population, the IAE model aims to estimate the rating of an individual user, considering their individual biases and inconsistencies. Therefore, the regressor learns the difference between the general population and the individual user, which is crucial for the IAE model.

Due to the limited number of 190 complete trials, the leave-one-subject-out cross-validation enables the use of the largest possible training data share. This approach involves training a separate regressor for each participant, utilizing all trials except those recorded from that specific person. On average, around seven trials per person are available for training, resulting in approximately 183 samples on average. The training process employs the variables identified in the previous chapter, excluding the estimated quality scores to solely base the model's estimation on individuality data. Thus, each model is trained using about 183 samples with 24 variables. Notably, none of the *Experience Ratings* variables are included in the training data. A separate train/test procedure is performed for each of the five QoE dimensions of interest.

The regressor employed is a Gradient Boosting Regressor (GBR) (Friedman, 2001) from the scikit-learn toolkit (Pedregosa et al., 2011) (version 1.2.1). GBR is an ensemble machine learning algorithm that iteratively combines weak prediction models, typically decision trees, to make accurate predictions in regression tasks. It is particularly effective in capturing nonlinear relationships in the data. Prior to training, the data are preprocessed by standardizing all features, removing the mean, and scaling to unit variance (all features have a mean of 0 and a variance of 1).

The GBR is configured with a maximum depth of nine for individual regression estimators, 350 boosting stages, and a learning rate of 0.01. These parameters were determined through an exploratory process, where the GBR outperformed other models such as linear regressors, support vector regressors, and multilayer perceptrons. Although overfitting is a common concern with small data sets, the leave-one-subject-out cross-validation helps mitigate this issue by ensuring that each estimator is only applied to unseen data. However, it is worth noting that better results may be achieved with larger data sets, as more data allow the model to learn and generalize more effectively.

7.1.5 Inferring Other Metrics Using Quality Estimations

As outlined in Sect. 4.2, I selected traditional quality dimensions, as well as effort and task load, as the QoE dimensions of interest. While NISQA and DOVER are specifically designed to estimate speech or video quality from datasets like the one used in this study, there are no models available that directly estimate effort and task load from conversation recordings. However, as shown in Chap. 5, Fig. 5.3, the quality of a call is highly correlated with effort and task load. Therefore, I propose using the same NISQA and DOVER scores used for quality estimation to also estimate effort and task load ratings. A similar procedure, but with a different dataset and signal-based estimator, was discussed in more detail in Spang et al. (2023a).

Since the estimated quality scores from the model cannot be directly associated with effort or task load, an Support Vector Regressor (SVR) is employed to map the quality score to the effort and task load ratings. An SVR aims to find a hyperplane in a high-dimensional feature space that maximally separates the training data points while minimizing the error between the predicted and actual output values. It accomplishes this by defining an ϵ-insensitive tube around the regression line, where deviations within this tube are considered acceptable errors. The algorithm strives to find the regression line that lies within this tube and has the maximum margin from the support vectors. The SVR uses an Radial Basis Function (RBF) kernel with coefficient $\gamma = 0.1$ and regularization parameters $C = 100$ and $\epsilon = 0.1$. The output of this preprocessing-regression yields the mapped score, which serves as the computationally estimated effort and task load and is subsequently compared to the individuality-informed estimation process described earlier for quality ratings.

7.2 Observations

The procedure outlined in Sect. 7.1.4 is applied to compare estimations from instrumental quality estimation models against the individuality-assisted approach. The following two sections present the results of this comparison, the first for quality-related experience dimensions and the second for effort and task load estimation. The final section of this chapter estimates all five QoE dimensions without any signal-based estimations, solely relying on individuality data. This

Table 7.1 Estimation accuracy performance metrics, comparing the instrumental estimation models against the IAE approach

Approach	SROCC	PLCC	RMSE	MAE	$adj.R^2$	R^2
Estimating Audiovisual Quality						
Baseline model	0.843	0.843	0.858	0.711	0.711	0.710
IAE	0.853	0.857	0.829	0.665	0.693	0.731
Estimating Speech Quality						
Baseline	0.768	0.810	0.982	0.770	0.655	0.655
IAE	0.801	0.823	0.962	0.750	0.623	0.669
Estimating Video Quality						
Baseline	0.809	0.843	0.840	0.636	0.711	0.711
IAE	0.827	0.842	0.853	0.637	0.660	0.702
Estimating Effort						
Baseline	0.805	0.798	13.746	10.637	0.630	0.630
IAE	0.798	0.787	14.111	10.731	0.556	0.610
Estimating Task Load						
Baseline	0.366	0.358	3.203	2.482	0.117	0.117
IAE	0.436	0.408	3.195	2.474	0.000	0.122

additional analysis demonstrates the effectiveness of the assessed variables and constructs and provides insights into the amount of variance that can be derived solely from studying the individual rating a given scenario.

7.2.1 *Quality Estimation*

The comparison between the signal-based **audiovisual quality estimation** (combining speech and video scores to obtain an audiovisual score, as described in Sect. 2.1.2.5) and the IAE approach revealed a significant difference in the predicted estimations, as determined by a Wilcoxon signed-rank test ($T = 7083.0$, $p = 0.009$, median difference between pairs of estimated scores: −0.081). As shown in Table 7.1, all statistical metrics, except for the adjusted R-squared value, improved using the IAE approach. However, the $adj.R^2$ score is unlikely to favor the IAE approach as it penalizes a large number of predictor variables. In contrast, signal-based estimations involve a single variable (the estimated score), while the IAE approach is based on 24 different variables. Overall, the IAE approach enhances the accuracy of audiovisual quality estimation compared to the quality estimation model. The contrasts are visualized in Fig. 7.1.

Similarly, the IAE approach improved the metrics for **speech quality estimation**. The difference between the estimated quality scores of the two approaches was statistically significant ($T = 5004.0$, $p < 0.001$, median difference between pairs of estimated scores: −0.139). However, the **video quality estimation** did not benefit from the addition of individuality information. The performance metrics favored the

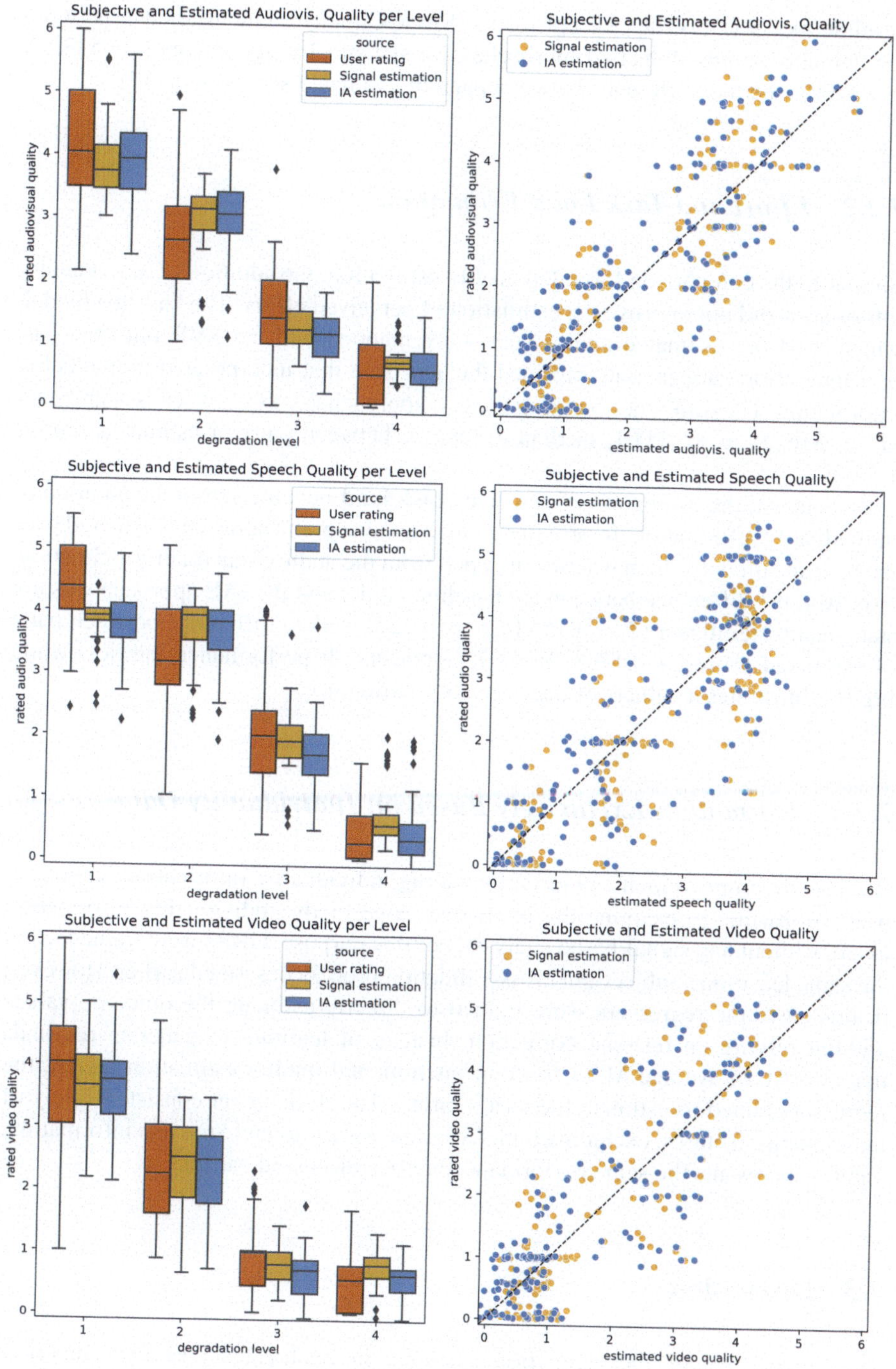

Fig. 7.1 The estimation performance was compared between a computational audiovisual quality estimation, the NISQA model, and the DOVER model. The left columns display box-plots comparing user ratings, signal-based estimations, and IAE for each experimental degradation condition. The right column shows scatter-plots depicting the fit between estimations and ratings

baseline model (DOVER), except for the SROCC metric. The difference between the estimated scores of the two approaches was statistically significant ($T = 5330.0$, $p < 0.001$, median difference between pairs of estimated scores: -0.157).

7.2.2 *Effort and Task Load Estimation*

Similar to the comparison for video quality estimation, the addition of individuality information did not improve the estimation of perceived **effort**. The baseline model, which used the estimated audiovisual score mapped with an SVR onto the participants' effort ratings, outperformed the approach that incorporated individuality information. The difference between the two approaches was statistically significant ($T = 5162.0$, $p < 0.001$, median difference between pairs of estimated scores: 1.591).

In contrast, the estimation of perceived **task load** benefited from the addition of individuality information, improving all metrics except the adjusted R-squared over the baseline model, which was also mapped from the audiovisual quality estimation. However, the difference between the baseline model and the IAE approach was not statistically significant ($T = 8155.0$, $p = 0.227$, median difference between pairs of estimated scores: -0.176). Table 7.1 presents all performance metrics, while Fig. 7.2 illustrates a collage of the estimation differences.

7.2.3 *Estimation Exclusively Based on Individuality Data*

Because the improvement achieved by adding individuality information is modest, it is worthwhile to examine the predictive value of the individuality information alone, without any signal-based analysis. For this purpose, GBRs were trained using the same leave-one-subject-out cross-validation process described earlier. However, in this case, the regressors were trained to directly estimate the outcome values without relying on residual correction. Instead of learning to generate residuals that correct an average MOS from an instrumental quality estimation model, the regressors aimed to estimate the user's rating. The training set consisted solely of individuality features, excluding both *Experience Ratings* and *Stimulus* information. Table 7.2 presents the same performance metrics discussed earlier.

7.3 Discussion

The comparison of the estimation accuracy in Sect. 7.2 shows only marginal improvements from the signal-based model estimation to the IAE. The accuracy increase for the estimation of audiovisual ratings from $PLCC = 0.843$ to $PLCC =$

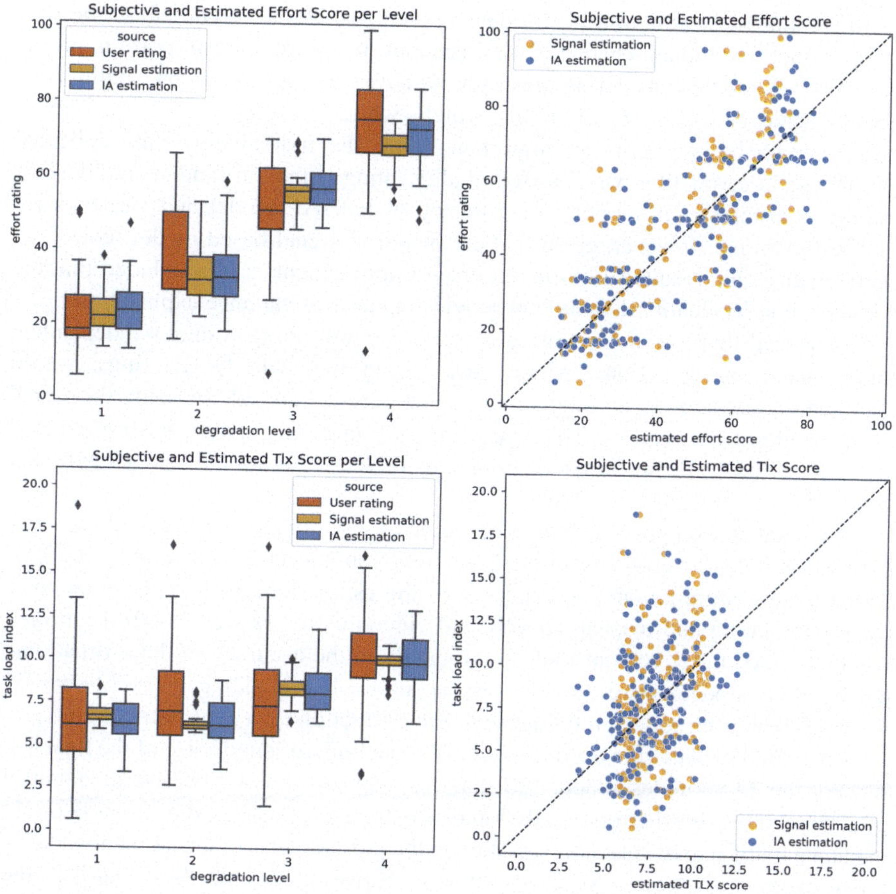

Fig. 7.2 Results of the estimation performance comparison between a computational effort estimation (top row) and task load estimation (bottom row) against an IAE

Table 7.2 Estimation accuracy performance metrics for the QoE estimator based exclusively on individuality data

Task	SROCC	PLCC	RMSE	MAE	$adj.R^2$	R^2
Estimating Audiovisual Quality	0.700	0.703	1.142	0.888	0.418	0.489
Estimating Speech Quality	0.671	0.675	1.236	0.994	0.379	0.454
Estimating Video Quality	0.665	0.635	1.209	0.884	0.319	0.402
Estimating Effort	0.571	0.587	18.307	14.828	0.253	0.344
Estimating Task Load	0.360	0.340	3.253	2.504	−0.037	0.090

0.857 seems minor, but considering the possible remaining margin for improvement (0.157), this increase of 0.014 represents 8.92% of the possible enhancement. This perspective offers a new understanding of the magnitude of the improvement.

Regarding the estimation performance solely based on individuality, the data suggest that individual characteristics account for a significant portion of the variability. A model considering only individuality variables to estimate audiovisual quality achieves a $PLCC$ of 0.703, which, while less precise than the signal-based approach, highlights an important point: the individuality data collected and its use in estimating personalized quality ratings function fundamentally. This finding is crucial as it supports the efficacy of the IAE model and variables for QoE estimations. However, given that the fusion of signal-based model scores and individuality data resulted in minimal or no improvement, it is concluded that the addition of individuality data did not provide significant variance explanation.

I theorized that the incorporation of variables describing stimulus-independent traits, states, and reactions, which have strong ties with QoE ratings, would significantly explain the missing variance that a model considering only the signal itself could not account for. This approach, as it currently stands, has not yielded satisfying results; the performance figures in Table 7.1 do not indicate a significant improvement in estimation accuracy.

The notable exception is the estimation of task load. Starting with a weak correlation between audiovisual quality estimation and task load ratings, the IAE added a substantial amount of accuracy to the estimation. Still, in the four other cases (estimating audiovisual, speech, and video quality, as well as effort ratings), where the IAE failed to contribute significantly to the accuracy of the estimations, the baseline models were already very accurate. Here, adding an additional 24 variables likely introduced more noise to the estimation than explanatory value.

A potential explanation for this outcome is the limited sample size of the final test set. For the 24 variables added, 190 samples may be too few to derive meaningful predictive value. Furthermore, individuality data and signal-based estimators might explain a lot of shared variance, resulting in the individuality information not adding much explanatory power to the estimation. However, despite these hurdles, the studies conducted, the analyses performed, and the model developed are valuable endeavors that provide a robust foundation for further research.

Chapter 8
Discussion, Future Work, and Conclusions

Science, my lad, is made up of mistakes, but they are mistakes which it is useful to make, because they lead little by little to the truth.
Jules Verne

This final chapter synthesizes the responses to the proposed research questions, discusses the pertinent findings, acknowledges the limitations of this work, and suggests potential directions for future research. Each significant result and decision from the preceding chapters are deliberated upon, concluding with an emphasis on the importance and practical applications of the work.

8.1 Answers to the Postulated Research Questions

In this section, I address the four research questions that were formulated in Sect. 2.4 one by one.

8.1.1 First Research Question: The Model

The first research question, regarding the model, is: How would a QoE formation process model look like, that focuses on individuality influences?

The model introduced in Chap. 6 meets the criteria of the research question. The original IAE model presented in Chap. 3 has been evaluated and analyzed through two different study paradigms and two different analysis methods. This has validated the proposed model, although it also necessitated some revisions.

R. P. Spang, *Individualized Quality of Experience Estimation in Audiovisual Communication*, T-Labs Series in Telecommunication Services,
https://doi.org/10.1007/978-3-031-67551-5_8

The lack of robustness of the "Dependent States" construct resulted in its removal, leading to a more reliable model structure. The final model in Fig. 6.3 demonstrates an evidence-backed framework that explains the influence of different individuality aspects on QoE ratings.

The SEM assigns weights to each model's edges, indicating the magnitude of relationships. Results confirm the traditional view that the strongest relationship exists between the stimulus and the QoE ratings, aligning with existing literature. However, this thesis aimed to explain the variance within these ratings, with the standardized weights between constructs providing insight into their relative importance. The behavioral reactions play a significant role, mainly manifested as conversational interactivity aspects. Frustration intolerance dimensions largely comprise the subsequent *Traits* construct. Preceding *Experience Ratings* and *Independent States* maintain a relationship with the *Experience Ratings*, encapsulating the QoE dimensions of interest.

8.1.2 Second Research Question: The Influences

The second research question, regarding the influences, is: Which individuality factors have the most dominant influence on QoE ratings?

The findings from Chaps. 5 and 6 answer this question by highlighting the most relevant variables impacting QoE ratings. Examples of these include, among others, the mean pauses between speaker changes, discomfort intolerance, the interlocutor relationship, or the preceding frustration rating.

The list of significant predictors from both studies offers a strong foundation for future investigations. Although some relationships failed to replicate in different study paradigms, exploring these anomalies could yield valuable insights for future research.

8.1.3 Third Research Question: The Performance

The third research question, regarding the performance, is: How relevant is an individuality-informed QoE estimation, as proposed by the developed model?

Chapter 7 scrutinizes the effectiveness and statistical benefit of an individuality enriched QoE estimation. The introduction of additional individuality data resulted in slight improvements in three out of five QoE dimensions. However, the increased effort and resources required to gather this additional information may not justify the marginal improvements. This trade-off is discussed further in Sect. 7.3.

8.1.4 Fourth Research Question: Viability of Simulations

The fourth research question, regarding the viability of simulations, is: Can a video-based conversation simulation be considered a reliable method to study human reactions to quality degradation?

The first half of Chap. 5 discusses the results of the conversation-simulation study, with 287 significant findings, only 33 of which could be replicated in a second study using a real conversation paradigm. Using both studies' data, I conclude that video-based conversations can effectively investigate quality changes' influences on emotional states, physiological and behavioral responses, as well as on QoE ratings.

However, it is crucial to validate these findings through diverse studies and study paradigms. Particularly, detecting subtle signals may prove more challenging in an interactive setting where participants engage in conversation. The analysis of various variables and constructs revealed the significance of interactivity dimensions in predicting QoE ratings. However, these dimensions cannot be obtained from a video-based conversation simulation. Consequently, while this paradigm may be valuable for other types of reactions, it would neglect the one that demonstrated the strongest relationships with the QoE dimensions.

Hence, while the paradigm is useful for pre-studies and early hypothesis testing, it is unlikely to replace conversation studies entirely. See also Sect. 8.3.

8.2 Assessment of Instruments and Variables

The prominence of the Big Five personality dimensions in existing literature contrasted with their rather marginal relevance within the present video-telephony studies. Although several significant relationships emerged in separate analyses for the two study paradigms (e.g., `effort` ~ `extraversion`, or `mental demand` ~ `openness`, see Table 5.2), only the "agreeableness" dimension showed consistent relationships in both datasets. Surprisingly, nearly all frustration intolerance dimensions, often overlooked in QoE research, demonstrated stable relationships across both the double-paradigm-validation approach and the indicator validation process of the SEM analysis. These results underscore the importance of incorporating the FDS in further QoE studies.

While recorded *Dependent States* add value to our dataset, their conceptualization as influences solely on subsequent trials may have undermined their predictive power. Since the IAE model aimed to estimate QoE dimensions rather than emotional states, warmth, or competence, these variables only played a role in preceding trials. Even though the relationship analyses revealed several such carryover effects, the final SEM favored a model variant without preceding *Dependent States*. Nonetheless, these variables enrich the dataset and make it valuable for research inquiries outside the strict QoE domain.

Employing physiological and behavioral measures depends heavily on the feature engineering conducted. A variety of features per sensor were extracted in this work, ranging from simple signal level averages to complex spectral power ratios (e.g., LF/HF ratio of HRV). A deeper investigation of these features is recommended when working with the recorded datasets, including deliberate data cleaning, particularly for the inherently noisy physiological signals.

Despite the challenges, the chosen selection of sensors and dimensions for investigation remains viable. EEG could offer further insights (Antons et al., 2012b; Engelke et al., 2016), albeit at the cost of added complexity to both the experimental process and the analysis pipeline.

8.3 Passive, Video-Based Conversation Simulation

The video simulation approach, designed to mimic a natural video-telephony situation via pre-recorded video clips, provided clean signals, particularly beneficial for motion-sensitive measurements such as EDA. Its controlled conditions, thanks to participants being less engaged compared to a real conversation, allowed us to identify 593 potentially relevant features. However, this method falls short of capturing the variability and different mental states introduced in an actual conversation, emphasizing the need to validate its findings in a conversational setting.

In Table 5.22, all variables demonstrating a significant relationship with the QoE dimensions across both study paradigms are listed. It is worth noting that among the three physiological and behavioral sensors utilized in both paradigms (EDA, Eye-Tracking, and Facial Configuration analyses), only the Eye-Tracking variables of mean and median fixation durations remained significant. This highlights the valuable and consistent nature of Eye-Tracking as a tool, as it consistently exhibits robust correlations with the QoE dimensions, even across different study paradigms.

With a retest of all 593 features in a conversation scenario, the robustness and significance of the findings were assured. Despite the inherently dynamic and noisier nature of a real conversation, a few significant effects emerged in this setting.

Responses to content-related questions and QoE ratings followed the pre-determined degradation levels closely, showcasing the effectiveness of the stimulus data set. The likeness of the stories varied slightly, while the actor was consistently appreciated, suggesting successful stimuli creation.

The compiled video dataset, including 160 video files and a descriptive table detailing computed quality scores, scripts, questions, and average user ratings, is publicly available (Spang et al., 2022b). Despite its limitations, the video simulation method, when validated in an actual conversation setting, contributes to a robust understanding of the IAE model of experience. Potential improvements to the dataset include extending its variability regarding actors, handling time warping and dropouts, and widening its relevance for researchers in different languages.

8.4 Conversation Experiment

The conversation simulation was followed by an evolution toward creating a real conversational scenario involving actual participant conversations. This required the development of a study platform that enabled a WebRTC connection between clients, traffic routing via a manipulated control server to introduce controlled network degradation, served questionnaires or instructions at appropriate times, logged all user input, and ensured experiment synchronization. On the sensor side, all devices logged their data in a synchronizable format with the timestamps and event markers of the call server. These software components enabled a semiautomatic experiment flow, necessary for managing two participants throughout the experiment's duration. The developed components themselves serve as tangible outcomes of this work, providing a ready-made platform for future conversational studies. What sets this platform apart from previous works is its notable modularity and seamless integration with third-party sensors.

To foster naturalistic conversations, participants performed two different tasks: story comparison and a survival task. Both tasks encouraged free turn-taking behavior and natural conversation progression. The aim was to test the IAE in a realistic scenario within a lab environment, contrasting the strictly controlled conversation-simulation approach. In retrospect, a middle-ground scenario involving an actual conversation with stricter controls might have yielded less noisy data and better results. Future experiments should consider selecting a more controlled, less flexible task, such as a random-number verification task, for example. Nevertheless, it is important to note that the interactivity dimensions, which solely represent the *Reactions* construct, would unlikely derive significant benefits from a more constrained experimental paradigm.

The dataset from the conversation experiments included 127 participants and contained 1140 trials. Like the conversation-simulation dataset, it is a valuable outcome of this work and can support further investigations.

8.5 Implications and Future Work

The possibility of separating the *Reactions* component of the IAE model into physiological and behavioral aspects could be considered as a means to maintain some physiological features. However, considering the final results, it might be beneficial to explore different preprocessing and analysis approaches entirely.

The studies carried out have generated a substantial dataset, opening opportunities for numerous analyses to further examine the relationships between individuality elements, reactions, and responses. For instance, questions surrounding the social relationship between interlocutors and its influence on their perception ratings can be addressed using this dataset. Additionally, physio-synchrony studies represent another interesting avenue for future research, among many others that could benefit from this dataset. Beyond secondary research questions, the primary domain of this work—the influence of individuality on QoE ratings—can be further explored.

Moreover, the methodologies and setups developed in this book hold significant potential for application in live monitoring environments. By leveraging the data collection mechanisms, including physiological and behavioral sensors, real-time feedback on QoE can be provided to clients during live video telephony sessions. The integration of tools such as ECG for heart rate variability, EDA for emotional arousal, and eye-tracking for cognitive load allows for a nuanced understanding of user states in real time. These insights can be used to dynamically adjust technical parameters such as encoding bitrate and packet loss to optimize the video and audio quality based on individual user responses. However, there is still more work to be done to refine these techniques for practical implementation. Future research should focus on developing robust algorithms and user interfaces that can facilitate seamless live monitoring and adjustments, ensuring higher satisfaction and engagement in real-world video telephony scenarios. By incorporating such live monitoring capabilities, the practical applications of this research can extend beyond theoretical contributions to offer tangible benefits in real-time client interactions.

The findings from Chap. 5 and Table 7.2, which involve the estimation results based solely on individuality, strongly suggest that individuality information accounts for a substantial share of variance. These results still support the belief that incorporating information about the rater could explain biases and inconsistencies previously considered random (Li et al., 2020; Tiotsop et al., 2023). However, another key finding of this work is that integrating all the relevant variables into a single estimator is not the most effective strategy. The results from all the analyses chapters suggest a need to further investigate the relationships between the variables, surpassing the scope of this thesis. For instance, exploring mediation and moderation effects across the wide array of variables could help to uncover the complex nature of the individuality relationships in future research.

8.6 Conclusion

Throughout the course of this research, I have delved deeply into the realm of Individuality Assisted Estimation (IAE), analyzing its potential to improve accuracy in QoE estimations. As demonstrated in this body of work, while there are indeed improvements in estimation accuracy, they are relatively marginal in comparison to the already high accuracy of signal-based estimators. This, however, doesn't diminish the value of the IAE approach; it highlights its potential utility in areas where the signal-based approach leaves room for improvement.

A fundamental takeaway from this research is the understanding that individuality data account for a significant share of the variability in estimation performance. This highlights that, while the signal-based approach is often more accurate, the inclusion of individuality variables, such as stimulus-independent traits, states, and reactions, can contribute to a more nuanced and personalized understanding of QoE ratings. This opens a new opportunity for future work and improvement in this field.

Interestingly, the IAE model demonstrated notable efficacy in the estimation of task load, outperforming the signal-based model. This finding underscores the possibility of a more targeted application of IAE where it can provide a better variance explanation than the signal-based approach. The relevance of this finding cannot be overstated as it indicates areas where the integration of additional variables can significantly enhance the accuracy of estimations.

Although the IAE approach failed to significantly enhance the accuracy of estimations in certain cases, it does not undermine the value of the methodology in any way. Instead, it points to the need for further investigation and refinement of the methodology. The possibly shared variance between individuality data and signal-based estimators, as well as the impact of sample size, are aspects that require further exploration. The recognition of these potential limitations and challenges only serves to broaden the scope for further work.

In the grand scheme of things, this research has yielded valuable insights. The methodologies developed, the extensive analyses conducted, and the wealth of data generated provide a firm foundation from which future investigations can leap. As the proposed methodology is further refined, the influence of individuality on QoE ratings will undoubtedly continue to be a compelling subject of inquiry. This research has set the stage for further investigation into the complex nature of individuality relationships and their implications on QoE ratings. As we move forward, the potential for significant advancements in this domain is promising.

Appendix A
Story Comparison Tasks

A.1 Story Comparison Tasks

As for the story comparison task, the narrative was exclusively displayed on the screen. It was visible for a duration of 2 minutes, during which participants were directed to memorize the story to the best of their ability. Following this period, the call began, and participants were not granted the opportunity for a second perusal of the story.

A.1.1 Story 1

A.1.1.1 Version A

Ein müder Student geht an einem Freitagnachmittag nach einem langen Tag in der Uni und mit vielen Vorlesungen endlich in den wohlverdienten Feierabend. Dabei nimmt er die U-Bahn der Linie U2 und muss nur 4 Stationen fahren. Während der Fahrt hat ihm seine Mutter per SMS darum gebeten, sich um das Abendessen zu kümmern, da es bei ihr auf Arbeit länger dauert. Er überlegt beim Lieblingsrestaurant der Mutter eine Pizza holen, doch dieses hat geschlossen, sodass er sich doch für einen Burger-Laden entschied. Zuhause angekommen bemerkt er einen Zettel vom Nachbarn im Briefkasten mit der Information, dass ein Paket, das an den Studenten adressiert war, fälschlicherweise zum Nachbarn geliefert wurde. Er klingelt bei besagtem Nachbarn und als das Päckchen erfolgreich den Besitzer gewechselt hatte, widmet sich der Student wieder dem Essen. Da die Mutter noch immer nicht heimgekommen war, aß er vor dem Fernsehgerät auf der Couch und schaute währenddessen eine Folge seiner Lieblingsserie. Der Student schlief kurze Zeit darauf ein und wurde wieder wach, als er seine Mutter hörte, wie diese nun am

R. P. Spang, *Individualized Quality of Experience Estimation in Audiovisual Communication*, T-Labs Series in Telecommunication Services,
https://doi.org/10.1007/978-3-031-67551-5

Küchentisch saß und sich über das Abendessen freute. Nach einem kurzen Gespräch gingen sie beide ins Bett.

A.1.1.2 Version B

Ein müder Student geht an einem Freitagnachmittag nach einem langen Tag in der Uni und mit vielen Vorlesungen endlich in den wohlverdienten Feierabend. Dabei nimmt er die U-Bahn der Linie U2 und muss bis fast zur Endstation fahren. Während der Fahrt hatte ihn seine Mutter angerufen und darum gebeten, sich um das Abendessen zu kümmern, da es bei ihr auf Arbeit länger dauert. Er überlegt beim Lieblingsrestaurant der Mutter eine Pizza holen, doch hatte er an diesem Abend mehr Lust auf einen Nudelauflauf statt Pizza. Zuhause angekommen bemerkt er einen Zettel vom Nachbarn im Briefkasten mit der Information, dass ein Paket, das an den Nachbarn adressiert war, fälschlicherweise zum Studenten geliefert wurde. Er klingelt bei besagtem Nachbarn und als das Päckchen erfolgreich den Besitzer gewechselt hatte, widmet sich der Student wieder dem Essen. Die Mutter kam nun auch nach Hause, aber da sie gern erst einmal ein Bad nehmen wollte, aß er vor dem Fernsehgerät auf der Couch und schaute währenddessen eine Folge seiner Lieblingsserie. Der Student schlief kurze Zeit darauf ein und wurde wieder wach, als er seine Mutter hörte, wie diese nun neben ihm auf dem Sofa Platz nahm und sich über das Abendessen freute. Nach einem kurzen Gespräch gingen sie beide ins Bett.

A.1.1.3 Supporting Questions

- An welchem Tag geht der Student nach Hause?
- Mit welcher Bahn fährt er und wie weit?
- Wie und mit wem kommuniziert er während der Fahrt?
- Warum wird kommuniziert?
- Was überlegt er sich zu essen zu holen?
- Was passiert mit dem Paket?
- Wo ist die Mutter als er anfängt zu essen?
- Wo isst der Student?
- Wo isst die Mutter?
- Was sieht sich der Student beim Essen an?

A.1.2 Story 2

A.1.2.1 Version A

Zwei Freundinnen trafen sich an einem sonnigen warmen Nachmittag im Frühling in einem Café, um sich nach längerer Zeit von ihren Urlaubserfahrungen zu erzählen. Sie bestellten je einen herzhaften Bagel mit Mozzarella und einen Kakao und begannen sich zu unterhalten. Die eine war für 3 ganze Wochen in die Berge gefahren und war viel Wandern. Sie hatte die Ruhe gern und konnte sich gut entspannen und fühlt sich ausgeruht. Die andere Freundin war stattdessen lieber eine Woche in die Stadt gefahren und hat viele Sehenswürdigkeiten besichtigt und war viel unterwegs. Doch auch sie fühlte sich ausgeruht und erholt. Nächstes Mal würde sie auch mal einen Urlaub in der Natur ausprobieren. Gemeinsam schmiedeten sie Pläne für zukünftige Ausflüge und planten auch, mal etwas zu zweit als Mädels-Trip zu unternehmen nächstes Jahr. Dieser Urlaub sollte eine Mischung aus Action und Entspannung werden. Sie würden auf jeden Fall eine Shoppingtour machen und am Sandstrand am Meer liegen. Mit diesen Worten verabschiedeten sie sich und machten sich wieder auf den Weg nach Hause.

A.1.2.2 Version B

Zwei Freundinnen trafen sich an einem regnerisch trüben kalten Nachmittag im Spätherbst in einem Café, um sich nach längerer Zeit von ihren Urlaubserfahrungen zu erzählen. Sie bestellten je eine leckere süße Waffel mit Vanillesoße und einen Kakao und begannen sich zu unterhalten. Die eine war für einen Kurzurlaub für 3 Tage in die Berge gefahren und war viel Wandern. Sie hatte die Ruhe gern und konnte sich gut entspannen und fühlt sich ausgeruht. Die andere Freundin war stattdessen lieber eine Woche in die Stadt gefahren und hat viele Sehenswürdigkeiten besichtigt und war viel unterwegs. Doch sie fühlte sich nicht wirklich erholt und es war sehr anstrengend. Nächstes Mal würde sie auch mal einen Urlaub in der Natur ausprobieren. Gemeinsam schmiedeten sie Pläne für zukünftige Ausflüge und planten auch, mal etwas zu viert mit ihren Partnern zu unternehmen nächstes Jahr. Dieser Urlaub sollte eine Mischung aus Action und Entspannung werden. Sie würden auf jeden Fall eine Shoppingtour machen und am beheizten Pool im Hotel liegen. Mit diesen Worten verabschiedeten sie sich und machten sich wieder auf den Weg nach Hause.

A.1.2.3 Supporting Questions

- Zu welcher Zeit treffen sich die beiden?
- Was wurde jeweils zu Essen und zu Trinken bestellt?
- Über was wollten sich die beiden unterhalten?
- Wie lange und wohin war die erste Freundin unterwegs?
- Wie empfand die erste Freundin den Urlaub?
- Wie lange und wohin war die andere Freundin unterwegs?
- Wie empfand die zweite Freundin den Urlaub?
- Was würde die Freundin beim nächsten Urlaub ausprobieren?
- Wer fährt mit in den nächsten Urlaub?
- Wie sieht der nächste geplante Urlaub aus?

A.1.3 Story 3

A.1.3.1 Version A

“Es freut mich Sie kennenzulernen”, sagte der Chef der Agentur und deutete dem Anwerber, sich auf einen sehr unbequem aussehenden Holzstuhl zu setzen. Der junge Mann ließ sich nieder und faltete die Hände im Schoß. Abwartend betrachtete er seinen potenziellen neuen Arbeitgeber, der in einem bunten freundlich aussehenden Hemd vor ihm stand und daraufhin das Job-Interview begann. In diesem Job würde der junge Mann als noch unterfahrener Junior-Manager einsteigen und die Hauptaufgabe würde darin bestehen, den Kundenkontakt zu leiten. Außerdem müssen viele Excel-Tabellen-Berechnungen durchgeführt werden, was dem jungen Mann aber gar nicht gefiel und womit er auch kaum Erfahrung hatte. Er sagte dem Chef ehrlich, dass er noch einiges lernen würde müssen, falls er hier anfangen würde. Der Chef nahm das zur Kenntnis, wirkte aber etwas zerknirscht und fuhr mit seinen Erzählungen fort. Am Ende fragte der Chef, ob es noch weitere Fragen gäbe. Der junge Mann nickte und erkundigte sich danach, wie lange es die Abteilung schon gab und wie das Team aufgebaut ist. Der Chef nannte ihm lächelnd die Zahlen und verabschiedete sich mit den Worten “Wir melden uns dann bei Ihnen!”.

A.1.3.2 Version B

“Es freut mich Sie kennenzulernen”, sagte der Chef der Agentur und deutete dem Anwerber, sich in einen sehr gemütlich und einladend aussehenden Sessel zu setzen. Der junge Mann ließ sich nieder und faltete die Hände im Schoß. Abwartend betrachtete er seinen potenziellen neuen Arbeitgeber, der in Hemd und formalen Jackett vor ihm stand und daraufhin das Job-Interview begann. In diesem Job würde der junge Mann als geübter Senior-Manager einsteigen und die Hauptaufgabe würde darin bestehen, den Kundenkontakt zu leiten. Außerdem müssen viele Excel-Tabellen-Berechnungen durchgeführt werden, was der junge Mann zwar erst selten gemacht hatte, sich das aber gut und gerne vorstellen könnte. Er sagte dem Chef ehrlich, dass er noch einiges lernen würde müssen, falls er hier anfangen würde. Der Chef nahm das zur Kenntnis, nicht sehr abgeneigt von dieser Aussage und fuhr mit seinen Erzählungen fort. Am Ende fragte der Chef, ob es noch weitere Fragen gäbe. Der junge Mann nickte und wie viele Mitarbeiter die Abteilung insgesamt hatte und wie das Team aufgebaut ist. Der Chef nannte ihm lächelnd die Zahlen und verabschiedete sich mit den Worten “Wir melden uns dann bei Ihnen!”.

A.1.3.3 Supporting Questions

- Wer leitet das Gespräch?
- Wie sind die Sitzmöbel im Büro, in dem das Gespräch stattfindet, ausgestattet?
- Was macht der Bewerber mit seinen Händen?

- Wie ist das Erscheinungsbild des Arbeitgebers?
- Als was bewirbt er sich und was sind seine Aufgaben?
- Wie reagiert er auf die Aufgaben?
- Wie kommuniziert er über seine Stärken und Schwächen?
- Wie reagiert der Chef darauf?
- Was ist die erste Frage, die der Bewerber stellt?
- Mit welchen Worten verabschiedet sich der Chef?

A.1.4 Story 4

A.1.4.1 Version A

In einem kleinen Süßwarenladen kaufte sich ein Mädchen auf dem Weg zur Schule fast jeden Tag etwas für die Hofpause. Heute kam es mal wieder in den Laden und berichtete dem Verkäufer, dass es diesmal sogar extra viel Taschengeld bekommen habe und sich etwas Besonderes aussuchen wolle. Der Verkäufer griff daraufhin in ein unscheinbares Regal hinter ihm und präsentierte stolz eine brandneue Sorte Schokolade, die es erst seit ein paar Tagen geben würde. Das Mädchen freute sich, hatte sich aber etwas anderes gewünscht. Es lief nun selbst im Laden herum und entdeckte eine bunte Tüte Bonbons mit Erdbeergeschmack. Begeistert legte es die Auswahl auf den Tresen und zählte die Münzen. Der Verkäufer hatte gute Laune und legte noch ein kleines Schokoladenbonbon als Extra dazu. In der Schule angekommen teilte das Mädchen die Süßigkeiten vollständig unter seinen Freunden auf und erzählte den anderen von der neuen Sorte, von der der Verkäufer erzählt hatte. Sie nahm sich vor, das nächste Mal eine dieser neuen Tafeln mitzubringen und sie ihren Klassenkameraden zu zeigen.

A.1.4.2 Version B

In einem kleinen Süßwarenladen kaufte sich ein Mädchen auf dem Weg zur Schule einmal im Monat etwas für die Hofpause. Heute kam es mal wieder in den Laden und berichtete dem Verkäufer, dass es diesmal nicht mehr so viel Taschengeld übrighatte und sich etwas Besonderes aussuchen wolle. Der Verkäufer griff daraufhin in eine auffällig beschriftete Kiste auf der Auslage und präsentierte stolz eine brandneue Sorte Schokolade, die es erst seit ein paar Tagen geben würde. Das Mädchen freute sich, hatte sich aber etwas anderes gewünscht. Es lief nun selbst im Laden herum und entdeckte eine bunte Tüte Bonbons mit Waldmeistergeschmack. Begeistert legte es die Auswahl auf den Tresen und zählte die Münzen. Der Verkäufer hatte gute Laune und legte noch ein kleines Schokoladenbonbon als Extra dazu. In der Schule angekommen gab das Mädchen ihren Freunden je ein Bonbon ab, behielt den Rest aber für sich und erzählte den anderen von der neuen Sorte, von der der Verkäufer erzählt hatte. Sie nahm sich vor, das nächste Mal mit ihren Klassenkameraden gemeinsam in den Laden zu gehen und die neue Schokolade zu probieren.

A.1.4.3 Supporting Questions

- Wie regelmäßig kommt das Mädchen im Laden vorbei?
- Wofür kauft es die Süßwaren?
- Was berichtet es dem Verkäufer?
- Was möchte das Mädchen sich deshalb aussuchen?
- Was zeigt der Verkäufer dem Mädchen und woher hat er es?
- Wie reagiert das Mädchen?
- Für was entscheidet sich das Mädchen stattdessen?
- Wie war die Laune des Verkäufers und was bekam das Mädchen noch dazu?
- Wie teilt es die Süßware in der Schule auf?
- Was sind die Pläne des Mädchens für den nächsten Besuch im Laden?

Appendix B
Scales and Questionnaires

B.1 ITU Quality Scale with Overflow in German

See Fig. B.1.

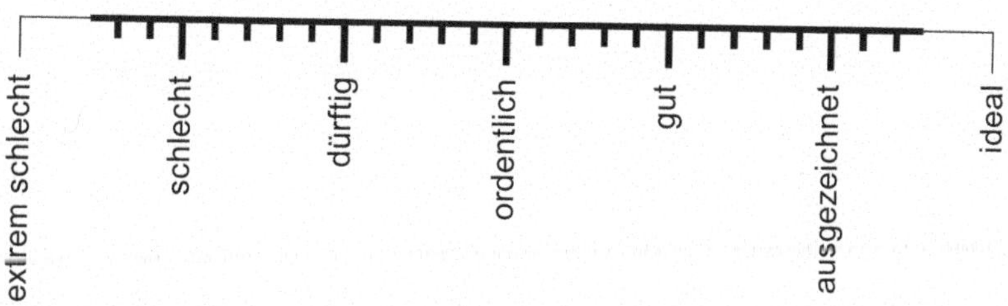

Fig. B.1 The ITU-T quality scale with overlow on both ends and German labels. This scale was used in all experiments to assess the audio, visual, and audiovisual quality of each trial

R. P. Spang, *Individualized Quality of Experience Estimation in Audiovisual Communication*, T-Labs Series in Telecommunication Services,
https://doi.org/10.1007/978-3-031-67551-5

Appendix C
Final Model Structure

See Fig. C.1.

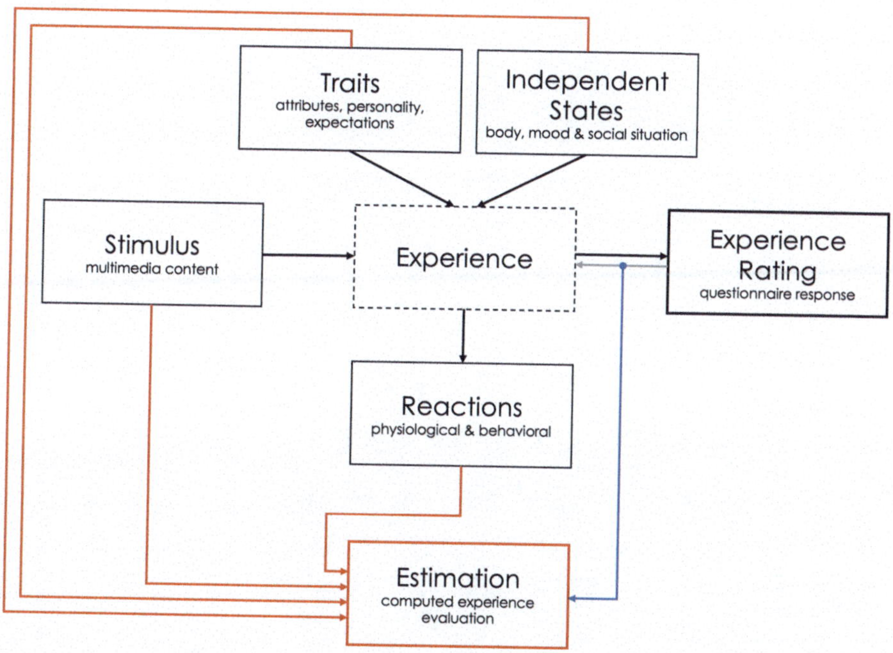

Fig. C.1 The updated, reduced model structure in the QoE estimation application. Black elements belong to the IAE model, orange lines are relationships between model entity variables with the estimator, the blue line is the relationship between preceding experience ratings (one trial before) and the estimation unit

R. P. Spang, *Individualized Quality of Experience Estimation in Audiovisual Communication*, T-Labs Series in Telecommunication Services,
https://doi.org/10.1007/978-3-031-67551-5

References

Abadi, M. K., Staiano, J., Cappelletti, A., Zancanaro, M., & Sebe, N. (2013). Multimodal engagement classification for affective cinema. In *2013 Humaine Association Conference on Affective Computing and Intelligent Interaction* (pp. 411–416). IEEE.
Ahern, S., & Beatty, J. (1979). Pupillary responses during information processing vary with scholastic aptitude test scores. *Science, 205*(4412), 1289–1292.
Airij, A. G., Sudirman, R., Sheikh, U. U., Khuan, L. Y., & Zakaria, N. A. (2020). Significance of electrodermal activity response in children with autism spectrum disorder. *Indonesian Journal of Electrical Engineering and Computer Science, 19*, 1113–1120.
Angrilli, A., Cherubini, P., Pavese, A., & Manfredini, S. (1997). The influence of affective factors on time perception. *Perception & Psychophysics, 59*, 972–982.
Antons, J.-N. (2015). *Neural correlates of quality perception for complex speech signals*. Springer.
Antons, J.-N., Arndt, S., Schleicher, R., & Möller, S. (2014). Brain activity correlates of quality of experience. *Quality of Experience: Advanced Concepts, Applications and Methods*, 109–119.
Antons, J.-N., Schleicher, R., Arndt, S., Möller, S., & Curio, G. (2012a). Too tired for calling? A physiological measure of fatigue caused by bandwidth limitations. In *2012 Fourth International Workshop on Quality of Multimedia Experience* (pp. 63–67). IEEE.
Antons, J.-N., Schleicher, R., Arndt, S., Moller, S., Porbadnigk, A. K., & Curio, G. (2012b). Analyzing speech quality perception using electroencephalography. *IEEE Journal of Selected Topics in Signal Processing, 6*(6), 721–731.
Apple Inc (2023). *Tracking and Visualizing Faces, blendShapes*. Apple Developer Documentation, accessed 2023-02-19. https://developer.apple.com/documentation/arkit/arfaceanchor/2928251-blendshapes.
Aranha, R. V., Nakamura, R., Tori, R., & Nunes, F. L. (2018). Personality traits impacts in virtual reality's user experience. In *2018 20th Symposium on Virtual and Augmented Reality (SVR)* (pp. 47–56). IEEE.
Arnau-González, P., Althobaiti, T., Katsigiannis, S., & Ramzan, N. (2017). Perceptual video quality evaluation by means of physiological signals. In *2017 Ninth International Conference on Quality of Multimedia Experience (QoMEX)* (pp. 1–6). IEEE.
Arndt, S. (2016). *Neural correlates of quality during perception of audiovisual stimuli*. Springer.
Arndt, S., Antons, J.-N., & Möller, S. (2014a). Is low quality media affecting the level of fatigue? In *2014 Sixth International Workshop on Quality of Multimedia Experience (QoMEX)* (pp. 47–48). IEEE.

R. P. Spang, *Individualized Quality of Experience Estimation in Audiovisual Communication*, T-Labs Series in Telecommunication Services,
https://doi.org/10.1007/978-3-031-67551-5

Arndt, S., Antons, J.-N., Schleicher, R., & Möller, S. (2016). Using electroencephalography to analyze sleepiness due to low-quality audiovisual stimuli. *Signal Processing: Image Communication, 42*, 120–129.

Arndt, S., Antons, J.-N., Schleicher, R., Möller, S., & Curio, G. (2012). Perception of low-quality videos analyzed by means of electroencephalography. In *2012 Fourth International Workshop on Quality of Multimedia Experience* (pp. 284–289). IEEE.

Arndt, S., Radun, J., Antons, J.-N., & Möller, S. (2014b). Using eye-tracking and correlates of brain activity to predict quality scores. In *2014 Sixth International Workshop on Quality of Multimedia Experience (QoMEX)* (pp. 281–285). IEEE.

Aron, A., Aron, E. N., & Smollan, D. (1992). Inclusion of other in the self scale and the structure of interpersonal closeness. *Journal of Personality and Social Psychology, 63*(4), 596.

Bampis, C. G., Li, Z., Moorthy, A. K., Katsavounidis, I., Aaron, A., & Bovik, A. C. (2017). Study of temporal effects on subjective video quality of experience. *IEEE Transactions on Image Processing, 26*(11), 5217–5231.

Baños, R. M., Etchemendy, E., Castilla, D., García-Palacios, A., Quero, S., & Botella, C. (2012). Positive mood induction procedures for virtual environments designed for elderly people. *Interacting with Computers, 24*(3), 131–138.

Barrett, L. F. (2017). *How emotions are made: The secret life of the brain*. Pan Macmillan.

Barrett, L. F., Adolphs, R., Marsella, S., Martinez, A. M., & Pollak, S. D. (2019). Emotional expressions reconsidered: Challenges to inferring emotion from human facial movements. *Psychological Science in the Public Interest, 20*(1), 1–68.

Barrett, L. F., & Bar, M. (2009). See it with feeling: Affective predictions during object perception. *Philosophical Transactions of the Royal Society B: Biological Sciences, 364*(1521), 1325–1334.

Beatty, J. (1982). Task-evoked pupillary responses, processing load, and the structure of processing resources. *Psychological Bulletin, 91*(2), 276.

Beatty, J., & Lucero-Wagoner, B. (2000). The pupillary system. In *Handbook of psychophysiology* (pp. 142–162). Cambridge University Press.

Belmudez, B. (2015). *Audiovisual quality assessment and prediction for videotelephony*. Springer.

Belmudez, B., Lewcio, B., & Möller, S. (2013). Call quality prediction for audiovisual time-varying impairments using simulated conversational structures. *Acta Acustica United with Acustica, 99*(5), 792–805.

Belmudez, B., & Möller, S. (2013). Audiovisual quality integration for interactive communications. *EURASIP Journal on Audio, Speech, and Music Processing, 2013*, 1–23.

Bernotat, J., & Eyssel, F. (2017). A robot at home—how affect, technology commitment, and personality traits influence user experience in an intelligent robotics apartment. In *2017 26th IEEE International Symposium on Robot and Human Interactive Communication (RO-MAN)* (pp. 641–646). IEEE.

Bland, J. M., & Altman, D. G. (1995). Multiple significance tests: The bonferroni method. *BMJ, 310*(6973), 170.

Bodden, M., & Jekosch, U. (1996). Entwicklung und Durchführung von Tests mit Versuchspersonen zur Verifizierung von Modellen zur Berechnung der Sprachübertragungsqualität. *Final report to a project funded by Deutsche Telekom AG (unpublished), Institut für Kommunikationsakustik, Ruhr Universität, Bochum.*

Bolls, P. D., Lang, A., & Potter, R. F. (2001). The effects of message valence and listener arousal on attention, memory, and facial muscular responses to radio advertisements. *Communication Research, 28*(5), 627–651.

Boucsein, W. (2012). *Electrodermal activity*. Springer Science & Business Media.

Bregman, A. S. (1994). *Auditory scene analysis: The perceptual organization of sound*. MIT Press.

Brooks, P., & Hestnes, B. (2010). User measures of quality of experience: Why being objective and quantitative is important. *IEEE Network, 24*(2), 8–13.

Cabin, R. J., & Mitchell, R. J. (2000). To bonferroni or not to Bonferroni: When and how are the questions. *Bulletin of the Ecological Society of America, 81*(3), 246–248.

Carroll, J. B., et al. (1993). *Human cognitive abilities: A survey of factor-analytic studies*. Number 1. Cambridge University Press.

Champseix, R., Ribiere, L., & Le Couedic, C. (2021). A python package for heart rate variability analysis and signal preprocessing. *Journal of Open Research Software, 9*(1).

Chandler, P., & Sweller, J. (1991). Cognitive load theory and the format of instruction. *Cognition and Instruction, 8*(4), 293–332.

Chun, M. M., & Wolfe, J. M. (2008). Visual attention. In *Blackwell handbook of sensation and perception* (pp. 272–310). Wiley Online Library.

Cohen, J. (2013). *Statistical power analysis for the behavioral sciences*. Academic Press.

Costa, P., & McCrae, R. (1989). Neo five-factor inventory (neo-ffi). *Odessa, FL: Psychological Assessment Resources*, 3.

Costa Jr, P. T., & McCrae, R. R. (1992). Four ways five factors are basic. *Personality and Individual Differences, 13*(6), 653–665.

Côté, N. (2011). *Integral and diagnostic intrusive prediction of speech quality*. Springer Science & Business Media.

Côté, N., Gautier-Turbin, V., & Möller, S. (2007). Influence of loudness level on the overall quality of transmitted speech. In *Audio Engineering Society Convention 123*. Audio Engineering Society.

Craig, S., D'Mello, S., Gholson, B., Witherspoon, A., Sullins, J., & Graesser, A. (2004). Emotions during learning: The first steps toward an affect sensitive intelligent tutoring system. In *E-Learn: World Conference on E-Learning in Corporate, Government, Healthcare, and Higher Education* (pp. 264–268). Association for the Advancement of Computing in Education (AACE).

Craig, S. D., D'Mello, S., Witherspoon, A., & Graesser, A. (2008). Emote aloud during learning with autotutor: Applying the facial action coding system to cognitive–affective states during learning. *Cognition and Emotion, 22*(5), 777–788.

Cuddy, A. J., Fiske, S. T., & Glick, P. (2008). Warmth and competence as universal dimensions of social perception: The stereotype content model and the bias map. *Advances in Experimental Social Psychology, 40*, 61–149.

Daniel, W. W. (1990). Friedman two-way analysis of variance by ranks. *Applied Nonparametric Statistics*, 262–274.

Danks, N. P., Sharma, P. N., & Sarstedt, M. (2020). Model selection uncertainty and multimodel inference in partial least squares structural equation modeling (PLS-SEM). *Journal of Business Research, 113*, 13–24.

De Moor, K., Mazza, F., Hupont, I., Quintero, M. R., Mäki, T., & Varela, M. (2014). Chamber QoE: A multi-instrumental approach to explore affective aspects in relation to quality of experience. In *Human Vision and Electronic Imaging XIX* (Vol. 9014, pp. 204–217). SPIE.

Démuth, A. (2013). Perception theories. *Kraków: Trnavská univerzita*, 2549.

Dijkstra, T. K. (2009). Latent variables and indices: Herman wold's basic design and partial least squares. In *Handbook of partial least squares: Concepts, methods and applications* (pp. 23–46). Springer.

Dijkstra, T. K., & Henseler, J. (2015). Consistent partial least squares path modeling. *MIS Quarterly, 39*(2), 297–316.

Dinas, P. C., Koutedakis, Y., & Flouris, A. D. (2013). Effects of active and passive tobacco cigarette smoking on heart rate variability. *International Journal of Cardiology, 163*(2), 109–115.

D'mello, S. K., & Graesser, A. (2010). Multimodal semi-automated affect detection from conversational cues, gross body language, and facial features. *User Modeling and User-Adapted Interaction, 20*, 147–187.

Dolan, R. J. (2002). Emotion, cognition, and behavior. *Science, 298*(5596), 1191–1194.

Donley, J., Ritz, C., & Shujau, M. (2014). Analysing the quality of experience of multisensory media from measurements of physiological responses. In *2014 Sixth International Workshop on Quality of Multimedia Experience (QoMEX)* (pp. 286–291). IEEE.

Döring, N., Moor, K. D., Fiedler, M., Schoenenberg, K., & Raake, A. (2022). Videoconference fatigue: A conceptual analysis. *International Journal of Environmental Research and Public Health, 19*(4), 2061.

Duchowski, A. T., Krejtz, K., Krejtz, I., Biele, C., Niedzielska, A., Kiefer, P., Raubal, M., & Giannopoulos, I. (2018). The index of pupillary activity: Measuring cognitive load vis-à-vis task difficulty with pupil oscillation. In *Proceedings of the 2018 CHI Conference on Human Factors in Computing Systems* (pp. 1–13).

Egan, D., Brennan, S., Barrett, J., Qiao, Y., Timmerer, C., & Murray, N. (2016). An evaluation of heart rate and electrodermal activity as an objective QoE evaluation method for immersive virtual reality environments. In *2016 Eighth International Conference on Quality of Multimedia Experience (QoMEX)* (pp. 1–6). IEEE.

Engeldrum, P. G. (2000). *Psychometric scaling: A toolkit for imaging systems development*. Imcotek Press.

Engelke, U., Darcy, D. P., Mulliken, G. H., Bosse, S., Martini, M. G., Arndt, S., Antons, J.-N., Chan, K. Y., Ramzan, N., & Brunnström, K. (2016). Psychophysiology-based QoE assessment: A survey. *IEEE Journal of Selected Topics in Signal Processing, 11*(1), 6–21.

ETSI Technical Report 102 506 (2007). *Speech Processing, Transmission and Quality Aspects (STQ); Estimating Speech Quality Per Call*. European Telecommunications Standards Institute, Sophia Antipolis, France.

Farias, M., Carli, M., & Mitra, S. (2005). Objective video quality metric based on data hiding. *IEEE Transactions on Consumer Electronics, 51*(3), 983–992.

Farias, M. C. (2010). Video quality metrics. *Digital Video*, 978–953.

Fedorikhin, A., & Cole, C. A. (2004). Mood effects on attitudes, perceived risk and choice: Moderators and mediators. *Journal of Consumer Psychology, 14*(1–2), 2–12.

Fiske, S. T. (1993). Controlling other people: The impact of power on stereotyping. *American Psychologist, 48*(6), 621.

Fiske, S. T., Cuddy, A. J., Glick, P., & Xu, J. (2002). A model of (often mixed) stereotype content: Competence and warmth respectively follow from perceived status and competition. *Journal of Personality and Social Psychology, 82*(6), 878.

Flägel, K., Galler, B., Steinhäuser, J., & Götz, K. (2019). The "national aeronautics and space administration-task load index"(nasa-tlx)-an instrument for measuring consultation workload within general practice: Evaluation of psychometric properties. *Zeitschrift fur Evidenz, Fortbildung und Qualitat im Gesundheitswesen, 147*, 90–96.

Forgas, J. P., & Bower, G. H. (1987). Mood effects on person-perception judgments. *Journal of Personality and Social Psychology, 53*(1), 53.

Friedman, J. H. (2001). Greedy function approximation: A gradient boosting machine. *Annals of Statistics*, 1189–1232.

Frijda, N. H. (1994). Varieties of affect: Emotions and episodes, moods, and sentiments. In *The nature of emotions: Fundamental questions*. Oxford University Press.

Furedy, J. J. (1983). Operational, analogical and genuine definitions of psychophysiology. *International Journal of Psychophysiology, 1*(1), 13–19.

Galloso, I., Palacios, J. F., Feijóo, C., & Santamaría, A. (2016). On the influence of individual characteristics and personality traits on the user experience with multi-sensorial media: An experimental insight. *Multimedia Tools and Applications, 75*, 12365–12408.

Giannakakis, G., Grigoriadis, D., Giannakaki, K., Simantiraki, O., Roniotis, A., & Tsiknakis, M. (2019). Review on psychological stress detection using biosignals. *IEEE Transactions on Affective Computing, 13*(1), 440–460.

Girard, J. M., Cohn, J. F., Jeni, L. A., Sayette, M. A., & De la Torre, F. (2015). Spontaneous facial expression in unscripted social interactions can be measured automatically. *Behavior Research Methods, 47*, 1136–1147.

Goldberg, L. R. (1990). An alternative "description of personality": The big-five factor structure. *Journal of Personality and Social Psychology, 59*(6), 1216.

Goldstein, E. B., & Cacciamani, L. (2021). *Sensation and perception*. Cengage Learning.

Gupta, R., Banville, H. J., & Falk, T. H. (2015). PhySyQX: A database for physiological evaluation of synthesised speech quality-of-experience. In *2015 IEEE workshop on applications of signal processing to audio and acoustics (WASPAA)* (pp. 1–5). IEEE.

Hair, J., Hult, G. T. M., Ringle, C., & Sarstedt, M. (2022). *A primer on partial least squares structural equation modeling (PLS-SEM).*

Hair Jr, J. F., Hult, G. T. M., Ringle, C. M., Sarstedt, M., Danks, N. P., & Ray, S. (2021). *Partial least squares structural equation modeling (PLS-SEM) using R: A workbook.* Springer Nature.

Hammer, F. (2006). *Quality aspects of packet-based interactive speech communication.* Forschungszentrum Telekommunikation Wien.

Hammer, F., & Reichl, P. (2005). How to measure interactivity in telecommunications. In *44th FITCE Congress*. Citeseer.

Hammer, F., Reichl, P., & Raake, A. (2005). The well-tempered conversation: Interactivity, delay and perceptual voip quality. In *IEEE International Conference on Communications, 2005. ICC 2005. 2005* (Vol. 1, pp. 244–249). IEEE.

Hands, D. S., & Avons, S. (2001). Recency and duration neglect in subjective assessment of television picture quality. *Applied Cognitive Psychology: The Official Journal of the Society for Applied Research in Memory and Cognition, 15*(6), 639–657.

Harrington, N. (2005). The frustration discomfort scale: Development and psychometric properties. *Clinical Psychology & Psychotherapy: An International Journal of Theory & Practice, 12*(5), 374–387.

Harrison, L., Skau, D., Franconeri, S., Lu, A., & Chang, R. (2013). Influencing visual judgment through affective priming. In *Proceedings of the SIGCHI Conference on Human Factors in Computing Systems* (pp. 2949–2958)

Hart, S. G., & Staveland, L. E. (1988). Development of NASA-TLX (task load index): Results of empirical and theoretical research. *Advances in Psychology, 52*, 139–183.

Henseler, J., Ringle, C. M., & Sarstedt, M. (2015). A new criterion for assessing discriminant validity in variance-based structural equation modeling. *Journal of the Academy of Marketing Science, 43*, 115–135.

Hess, E. H., & Polt, J. M. (1964). Pupil size in relation to mental activity during simple problem-solving. *Science, 143*(3611), 1190–1192.

Hofstede, G. (1984). *Culture's consequences: International differences in work-related values* (Vol. 5). Sage.

Hofstede, G. (2011). Dimensionalizing cultures: The hofstede model in context. *Online Readings in Psychology and Culture, 2*(1), 2307–0919.

Holland, J. L. (1997). *Making vocational choices: A theory of vocational personalities and work environments*. Psychological Assessment Resources.

Holm, S. (1979). A simple sequentially rejective multiple test procedure. *Scandinavian Journal of Statistics*, 65–70.

Hulland, J. (1999). Use of partial least squares (PLS) in strategic management research: A review of four recent studies. *Strategic Management Journal, 20*(2), 195–204.

ISO Standard 9000 (2000). *Quality management systems: Fundamentals and vocabulary*. International Organization for Standardization.

ITU-T Contribution COM 12-61-E (1998). *Study of the influence of experimental context on the relationship between audio, video and audiovisual subjective qualities. Technical report.* International Telecommunication Union.

ITU-T Contribution COM 12-64-E (1997). *Relations between audio, video and audiovisual quality. Technical report.* International Telecommunication Union.

ITU-T Contribution COM 12-64-E (1998). *Results of an audiovisual desktop video teleconferencing subjective experiment. Technical report.* International Telecommunication Union.

ITU-T Recommendation P.10 (2017). *Vocabulary for performance, quality of service and quality of experience.* International Telecommunication Union.

ITU-T Recommendation P.1301 (2017). *Subjective quality evaluation of audio and audiovisual multiparty telemeetings.* International Telecommunication Union.

ITU-T Recommendation P.1401 (2020). *Methods, metrics and procedures for statistical evaluation, qualification and comparison of objective quality prediction models*. International Telecommunication Union.
ITU-T Recommendation P.563 (2004). *Single-ended method for objective speech quality assessment in narrow-band telephony applications*. International Telecommunication Union.
ITU-T Recommendation P.800 (1996). *Methods for subjective determination of transmission quality*. International Telecommunication Union.
ITU-T Recommendation P.804 (2017). *Subjective diagnostic test method for conversational speech quality analysis*. International Telecommunication Union.
ITU-T Recommendation P.805 (2007). *Subjective evaluation of conversational quality*. International Telecommunication Union.
ITU-T Recommendation P.911 (1998). *Subjective audiovisual quality assessment methods for multimedia applications*. International Telecommunication Union.
ITU-T Recommendation P.918 (2020). *Dimension-based subjective quality evaluation for video content*. International Telecommunication Union.
ITU-T Recommendation P.920 (2000). *Interactive test methods for audiovisual communications*. International Telecommunication Union.
Jekosch, U. (2005). *Voice and speech quality perception: assessment and evaluation*. Springer Science & Business Media.
Jöreskog, K. G. (1971). Simultaneous factor analysis in several populations. *Psychometrika, 36*(4), 409–426.
Jumisko-Pyykkö, S. (2011). *User-centered quality of experience and its evaluation methods for mobile television*. Dissertation, Tampere University of Technology.
Kahneman, D. (2011). *Thinking, fast and slow*. Macmillan.
Kahneman, D., & Beatty, J. (1966). Pupil diameter and load on memory. *Science, 154*(3756), 1583–1585.
Kaplan, S. (1995). The restorative benefits of nature: Toward an integrative framework. *Journal of Environmental Psychology, 15*(3), 169–182.
Killermann, S. (2018). The genderbread person version 4. *It's pronounced metrosexual*.
Kim, H.-G., Cheon, E.-J., Bai, D.-S., Lee, Y. H., & Koo, B.-H. (2018). Stress and heart rate variability: A meta-analysis and review of the literature. *Psychiatry Investigation, 15*(3), 235.
Kiousis, S. (2002). Interactivity: A concept explication. *New Media & Society, 4*(3), 355–383.
Kleiger, R. E., Stein, P. K., Bigger Jr, J. T. (2005). Heart rate variability: Measurement and clinical utility. *Annals of Noninvasive Electrocardiology, 10*(1), 88–101.
Kołodziej, M., Tarnowski, P., Majkowski, A., & Rak, R. (2019). Electrodermal activity measurements for detection of emotional arousal. *Bulletin of the Polish Academy of Sciences. Technical Sciences, 67*(4), 813–826.
Köster, F. (2017). *Multidimensional analysis of conversational telephone speech*. Springer.
Köster, F., Guse, D., Wältermann, M., & Möller, S. (2015). Comparison between the discrete ACR scale and an extended continuous scale for the quality assessment of transmitted speech. *Fortschritte der Akustik, DAGA*.
Köster, F., & Möller, S. (2014). Analyzing perceptual dimensions of conversational speech quality. In *Proceedings of INTERSPEECH* (pp. 2041–2045).
Krueger, M., Schulte, M., Zokoll, M. A., Wagener, K. C., Meis, M., Brand, T., & Holube, I. (2017). Relation between listening effort and speech intelligibility in noise. *American Journal of Audiology, 26*(3S), 378–392.
Lang, A. (2000). The limited capacity model of mediated message processing. *Journal of Communication, 50*(1), 46–70.
Lewinski, P. (2015). Automated facial coding software outperforms people in recognizing neutral faces as neutral from standardized datasets. *Frontiers in Psychology, 6*, 1386.
Li, Z., Bampis, C. G., Krasula, L., Janowski, L., & Katsavounidis, I. (2020). A simple model for subject behavior in subjective experiments. Preprint. arXiv:2004.02067.

Makowski, D., Pham, T., Lau, Z. J., Brammer, J. C., Lespinasse, F., Pham, H., Schölzel, C., & Chen, S. H. A. (2021). NeuroKit2: A python toolbox for neurophysiological signal processing. *Behavior Research Methods, 53*(4), 1689–1696.

Markus, H. R., & Kitayama, S. (1991). Culture and the self: Implications for cognition, emotion, and motivation. *Psychological Review, 98*(2), 224.

Martinez, H., Farias, M. C., & Hines, A. (2019). Navidad: A no-reference audio-visual quality metric based on a deep autoencoder. In *2019 27th European Signal Processing Conference (EUSIPCO)* (pp. 1–5). IEEE.

Mather, G. (2014). *Essentials of sensation and perception*. Routledge.

McAdams, D. P. (2001). The psychology of life stories. *Review of General Psychology, 5*(2), 100–122.

McDaniel, B., D'Mello, S., King, B., Chipman, P., Tapp, K., & Graesser, A. (2007). Facial features for affective state detection in learning environments. In *Proceedings of the Annual Meeting of the Cognitive Science Society* (Vol. 29).

Mischel, W., & Shoda, Y. (1995). A cognitive-affective system theory of personality: Reconceptualizing situations, dispositions, dynamics, and invariance in personality structure. *Psychological Review, 102*(2), 246.

Mittag, G., & Möller, S. (2019). Non-intrusive speech quality assessment for super-wideband speech communication networks. In *ICASSP 2019-2019 IEEE International Conference on Acoustics, Speech and Signal Processing (ICASSP)* (pp. 7125–7129). IEEE.

Mittag, G., Naderi, B., Chehadi, A., & Möller, S. (2021). NISQA: A deep CNN-self-attention model for multidimensional speech quality prediction with crowdsourced datasets. Preprint. arXiv:2104.09494.

Möller, S., Engelbrecht, K.-P., Kühnel, C., Wechsung, I., & Weiss, B. (2009). A taxonomy of quality of service and quality of experience of multimodal human-machine interaction. In *2009 International Workshop on Quality of Multimedia Experience* (pp. 7–12). IEEE.

Montalvo, S., Martinez, A., Arias, S., Lozano, A., Gonzalez, M. P., Dietze-Hermosa, M. S., Boyea, B. L., & Dorgo, S. (2022). Commercial smart watches and heart rate monitors: A concurrent validity analysis. *The Journal of Strength & Conditioning Research*, 10–1519.

Moore, B. C. (2012). *An introduction to the psychology of hearing*. Brill.

Moore, M. P., & Martin, R. A. (2019). On the evolution of carry-over effects. *Journal of Animal Ecology, 88*(12), 1832–1844.

Morris, M. G., & Venkatesh, V. (2000). Age differences in technology adoption decisions: Implications for a changing work force. *Personnel Psychology, 53*(2), 375–403.

Mrvelj, Š., & Matulin, M. (2021). Modeling the level of user frustration for the impaired telemeeting service using user frustration susceptibility index (UFSI). *Electronics, 10*(18), 2202.

Newman, E. J., & Schwarz, N. (2018). Good sound, good research: How audio quality influences perceptions of the research and researcher. *Science Communication, 40*(2), 246–257.

Oliver, R. L. (1980). A cognitive model of the antecedents and consequences of satisfaction decisions. *Journal of Marketing Research, 17*(4), 460–469.

Pedregosa, F., Varoquaux, G., Gramfort, A., Michel, V., Thirion, B., Grisel, O., Blondel, M., Prettenhofer, P., Weiss, R., Dubourg, V., Vanderplas, J., Passos, A., Cournapeau, D., Brucher, M., Perrot, M., & Duchesnay, E. (2011). Scikit-learn: Machine learning in Python. *Journal of Machine Learning Research, 12*, 2825–2830.

Pilcher, J. J., & Huffcutt, A. I. (1996). Effects of sleep deprivation on performance: A meta-analysis. *Sleep, 19*(4), 318–326.

Plomin, R., & Daniels, D. (1987). Why are children in the same family so different from one another? *Behavioral and Brain Sciences, 10*(1), 1–16.

Pornpongtechavanich, P., & Daengsi, T. (2019). Video telephony-quality of experience: A simple qoe model to assess video calls using subjective approach. *Multimedia Tools and Applications, 78*, 31987–32006.

Powers, S. R., Rauh, C., Henning, R. A., Buck, R. W., & West, T. V. (2011). The effect of video feedback delay on frustration and emotion communication accuracy. *Computers in Human Behavior, 27*(5), 1651–1657.

Privitera, A. J. (2018). Sensation and perception. *General Psychology, FA18*, 64.

Punsawad, Y., Aempedchr, S., Wongsawat, Y., & Panichkun, M. (2011). Weighted-frequency index for EEG-based mental fatigue alarm system. *International Journal of Applied Biomedical Engineering, 4*(1), 37.

Qualinet (2013). *Qualinet white paper on definitions of quality of experience*. European Network on Quality of Experience in Multimedia Systems and Services.

Raake, A. (2006). *Speech quality of VoIP: Assessment and prediction*. John Wiley & Sons.

Raake, A. (2007). *Speech quality of VoIP: Assessment and prediction*. John Wiley & Sons.

Raake, A., & Blauert, J. (2013). Comprehensive modeling of the formation process of sound-quality. In *2013 Fifth International Workshop on Quality of Multimedia Experience (QoMEX)* (pp. 76–81). IEEE.

Raake, A., & Egger, S. (2014). Quality and quality of experience. In *Quality of experience: Advanced concepts, applications and methods* (pp. 11–33). Springer.

Rauthmann, J. F., Sherman, R. A., & Funder, D. C. (2015). Principles of situation research: Towards a better understanding of psychological situations. *European Journal of Personality, 29*(3), 363–381.

Reichl, P., Egger, S., Möller, S., Kilkki, K., Fiedler, M., Hoßfeld, T., Tsiaras, C., & Asrese, A. (2015). Towards a comprehensive framework for qoe and user behavior modelling. In *2015 Seventh International Workshop on Quality of Multimedia Experience (QoMEX)* (pp. 1–6). IEEE.

Reichl, P., & Hammer, F. (2004). Hot discussion or frosty dialogue? Towards a temperature metric for conversational interactivity. In *Eighth International Conference on Spoken Language Processing*.

Reiter, U., Brunnström, K., De Moor, K., Larabi, M.-C., Pereira, M., Pinheiro, A., You, J., & Zgank, A. (2014). Factors influencing quality of experience. In *Quality of experience: Advanced concepts, applications and methods* (pp. 55–72). Springer.

Reiter, U., & You, J. (2010). Estimating perceived audiovisual and multimedia quality—A survey. In *IEEE International Symposium on Consumer Electronics (ISCE 2010)* (pp. 1–6). IEEE.

Rumpa, L. D., Wibawa, A. D., Attamimi, M., Sampelawang, P., Purnomo, M. H., & Palelleng, S. (2018). Analysis on human heart signal during sad video stimuli using heart rate variability triangular index (HRVi). In *2018 International Conference on Computer Engineering, Network and Intelligent Multimedia (CENIM)* (pp. 25–28). IEEE.

Russell, J. A. (1980). A circumplex model of affect. *Journal of Personality and Social Psychology, 39*(6), 1161.

Russell, J. A. (2003). Core affect and the psychological construction of emotion. *Psychological Review, 110*(1), 145.

Russell, J. A., Weiss, A., & Mendelsohn, G. A. (1989). Affect grid: A single-item scale of pleasure and arousal. *Journal of Personality and Social Psychology, 57*(3), 493.

Saffaryazdi, N., Goonesekera, Y., Saffaryazdi, N., Hailemariam, N. D., Temesgen, E. G., Nanayakkara, S., Broadbent, E., & Billinghurst, M. (2022). Emotion recognition in conversations using brain and physiological signals. In *27th International Conference on Intelligent User Interfaces* (pp. 229–242).

Schiffner, F., & Möller, S. (2016). Diving into perceptual space: Quality relevant dimensions for video telephony. In *2016 8th International Conference on Quality of Multimedia Experience (QoMEX 2016)*. IEEE-Signal Processing Society.

Schiffner, F., & Möller, S. (2017). Defining the relevant perceptual quality space for video and video-telephony. In *2017 Ninth International Conference on Quality of Multimedia Experience (QoMEX)* (pp. 1–3). IEEE.

Schmidt, S., Uhrig, S., & Reuschel, D. (2020). Investigating the relationship of mental immersion and physiological measures during cloud gaming. In *2020 Twelfth International Conference on Quality of Multimedia Experience (QoMEX)* (pp. 1–6). IEEE.

Schmitt, M., Redi, J., Bulterman, D., & Cesar, P. S. (2017). Towards individual QoE for multiparty videoconferencing. *IEEE Transactions on Multimedia, 20*(7), 1781–1795.

Schnabel, K., Asendorpf, J. B., & Ostendorf, F. (2002). Replicable types and subtypes of personality: German NEO-PI-R versus NEO-FFI. *European Journal of Personality, 16*(S1), S7–S24.

Scholler, S., Bosse, S., Treder, M. S., Blankertz, B., Curio, G., Muller, K.-R., & Wiegand, T. (2012). Toward a direct measure of video quality perception using EEG. *IEEE Transactions on Image Processing, 21*(5), 2619–2629.

Schwartz, S. H. (1992). Universals in the content and structure of values: Theoretical advances and empirical tests in 20 countries. In *Advances in Experimental Social Psychology* (Vol. 25, pp. 1–65). Elsevier.

Scott, M. J., Guntuku, S. C., Lin, W., & Ghinea, G. (2016). Do personality and culture influence perceived video quality and enjoyment? *IEEE Transactions on Multimedia, 18*(9), 1796–1807.

Sharma, P. N., Shmueli, G., Sarstedt, M., Danks, N., & Ray, S. (2021). Prediction-oriented model selection in partial least squares path modeling. *Decision Sciences, 52*(3), 567–607.

Sheikh, H. R., Sabir, M. F., & Bovik, A. C. (2006). A statistical evaluation of recent full reference image quality assessment algorithms. *IEEE Transactions on Image Processing, 15*(11), 3440–3451.

Sirois, S., & Brisson, J. (2014). Pupillometry. *Wiley Interdisciplinary Reviews: Cognitive Science, 5*(6), 679–692.

Skowronek, J., & Raake, A. (2015). Conceptual model of multiparty conferencing and telemeeting quality. In *2015 Seventh International Workshop on Quality of Multimedia Experience (QoMEX)* (pp. 1–6). IEEE.

Spang, R. P., El Hajal, K., Möller, S., & Cernak, M. (2023a). Personalized task load prediction in speech communication. In *ICASSP 2023-2023 IEEE International Conference on Acoustics, Speech and Signal Processing (ICASSP)* (pp. 1–5). IEEE.

Spang, R. P., Machačík, O., Pieper, K., Vergari, M., Voigt-Antons, J.-N. (2022a). Reconstruction and physiological basis of Samsung's Galaxy watch stress score. In *HCI International 2022 Posters: 24th International Conference on Human-Computer Interaction, HCII 2022, Virtual Event, June 26–July 1, 2022, Proceedings, Part IV* (pp. 445–452). Springer.

Spang, R. P., Voigt-Antons, J.-N., & Möller, S. (2022b). The storytime dataset: Simulated videotelephony clips for quality perception research. In *2022 Fourteenth International Conference on Quality of Multimedia Experience (QoMEX)* (pp. 1–6). IEEE.

Spang, R. P., Wardah, W., Schmitt, V., & Möller, S. (2023b). Comparing simulated and real conversations for qoe assessments: Insights from arkit-based facial configuration analyses. In *2023 Fifteenth International Conference on Quality of Multimedia Experience (QoMEX)* (pp. 1–4). IEEE.

Spang, R. P., Wardah, W., Schmitt, V., & Möller, S. (2023c). Unraveling the hangry rater: Nonlinear effects of hunger on multimedia quality perception. In *2023 Fifteenth International Conference on Quality of Multimedia Experience (QoMEX)* (pp. 1–4). IEEE.

Spiers, J. P., Silke, B., McDermott, U., Shanks, R. G., & Harron, D. W. (1993). Time and frequency domain assessment of heart rate variability: A theoretical and clinical appreciation. *Clinical Autonomic Research, 3*, 145–158.

Sweller, J. (1988). Cognitive load during problem solving: Effects on learning. *Cognitive Science, 12*(2), 257–285.

Tiotsop, L. F., Servetti, A., & Masala, E. (2023). A scoring model considering the variability of subjects' characteristics in subjective experiments. In *2023 Fifteenth International Conference on Quality of Multimedia Experience (QoMEX)* (pp. 1–6). IEEE.

Uhrig, S. (2021). *Human information processing in speech quality assessment*. Springer.

Ushiyama, T., Mizushige, K., Wakabayashi, H., Nakatsu, T., Ishimura, K., Tsuboi, Y., Maeta, H., & Suzuki, Y. (2008). Analysis of heart rate variability as an index of noncardiac surgical stress. *Heart and Vessels, 23*, 53–59.

Valkenburg, P. M., & Peter, J. (2013). The differential susceptibility to media effects model. *Journal of Communication, 63*(2), 221–243.

Venkatesh, V., Morris, M. G., Davis, G. B., & Davis, F. D. (2003). User acceptance of information technology: Toward a unified view. *MIS Quarterly*, 425–478.

Venkatesh, V., Thong, J. Y., & Xu, X. (2012). Consumer acceptance and use of information technology: Extending the unified theory of acceptance and use of technology. *MIS Quarterly*, 157–178.

von der Malsburg, T. (2019). Saccade and fixation detection in R. accessed 2023-03-05.

Wagner, J., Kim, J., & André, E. (2005). From physiological signals to emotions: Implementing and comparing selected methods for feature extraction and classification. In *2005 IEEE International Conference on Multimedia and Expo* (pp. 940–943). IEEE.

Wältermann, M. (2013). *Dimension-based quality modeling of transmitted speech*. Springer Science & Business Media.

Wältermann, M., Raake, A., & Möller, S. (2010). Quality dimensions of narrowband and wideband speech transmission. *Acta Acustica United with Acustica, 96*(6), 1090–1103.

Wang, Y., Ke, J., Talebi, H., Yim, J. G., Birkbeck, N., Adsumilli, B., Milanfar, P., & Yang, F. (2021). Rich features for perceptual quality assessment of UGC videos. In *Proceedings of the IEEE/CVF Conference on Computer Vision and Pattern Recognition* (pp. 13435–13444).

Wechsung, I., Engelbrecht, K.-P., Kühnel, C., Möller, S., & Weiss, B. (2012). Measuring the quality of service and quality of experience of multimodal human–machine interaction. *Journal on Multimodal User Interfaces, 6*, 73–85.

Weiss, B., Möller, S., Raake, A., Berger, J., & Ullmann, R. (2009). Modeling call quality for time-varying transmission characteristics using simulated conversational structures. *Acta Acustica United with Acustica, 95*(6), 1140–1151.

Wilcoxon, F. (1945). Individual comparisons by ranking methods. *Biometrics Bulletin, 1*(6), 80–83.

Wu, H., Chen, C., Hou, J., Liao, L., Wang, A., Sun, W., Yan, Q., & Lin, W. (2022a). FAST-VQA: Efficient end-to-end video quality assessment with fragment sampling. In *European Conference on Computer Vision* (pp. 538–554). Springer.

Wu, H., Liao, L., Chen, C., Hou, J., Wang, A., Sun, W., Yan, Q., & Lin, W. (2022b). Disentangling aesthetic and technical effects for video quality assessment of user generated content. Preprint. arXiv:2211.04894.

Yamagishi, K., & Hayashi, T. (2005). Analysis of psychological factors for quality assessment of interactive multimodal service. In *Human Vision and Electronic Imaging X* (Vol. 5666, pp. 130–138). SPIE.

Ying, Z., Pan, D., & Shi, P. (2023). Blind video quality assessment for ultra-high-definition video based on super-resolution and deep reinforcement learning. *Sensors, 23*(3), 1511.

You, J., Reiter, U., Hannuksela, M. M., Gabbouj, M., & Perkis, A. (2010). Perceptual-based quality assessment for audio–visual services: A survey. *Signal Processing: Image Communication, 25*(7), 482–501.

Zhang, L., Zhang, L., & Bovik, A. C. (2015). A feature-enriched completely blind image quality evaluator. *IEEE Transactions on Image Processing, 24*(8), 2579–2591.

Zheng, L. J., Mountstephens, J., & Teo, J. (2020). Four-class emotion classification in virtual reality using pupillometry. *Journal of Big Data, 7*, 1–9.

Index

R. P. Spang, *Individualized Quality of Experience Estimation in Audiovisual Communication*, T-Labs Series in Telecommunication Services,
https://doi.org/10.1007/978-3-031-67551-5

The manufacturer's authorised representative in the EU is Springer Nature Customer Service Centre GmbH, Europaplatz 3, 69115 Heidelberg, Germany. If you have any concerns regarding our products, please contact ProductSafety@springernature.com

Printed and bound by CPI Group (UK) Ltd, Croydon, CR0 4YY
23/09/2026
02221542-0017